T0281098

Cambridge Elements ☰

Elements in the Philosophy of Biology
edited by
Grant Ramsey
KU Leuven
Michael Ruse
Florida State University

THE MISSING TWO-THIRDS OF EVOLUTIONARY THEORY

Robert N. Brandon and Daniel W. McShea
Duke University

CAMBRIDGE
UNIVERSITY PRESS

CAMBRIDGE
UNIVERSITY PRESS

University Printing House, Cambridge CB2 8BS, United Kingdom

One Liberty Plaza, 20th Floor, New York, NY 10006, USA

477 Williamstown Road, Port Melbourne, VIC 3207, Australia

314–321, 3rd Floor, Plot 3, Splendor Forum, Jasola District Centre, New Delhi – 110025, India

79 Anson Road, #06–04/06, Singapore 079906

Cambridge University Press is part of the University of Cambridge.

It furthers the University's mission by disseminating knowledge in the pursuit of education, learning, and research at the highest international levels of excellence.

www.cambridge.org
Information on this title: www.cambridge.org/9781108716680
DOI: 10.1017/9781108591508

First published 2020

A catalogue record for this publication is available from the British Library.

ISBN 978-1-108-71668-0 Paperback
ISSN 2515-1126 (online)
ISSN 2515-1118 (print)

The Missing Two-Thirds of Evolutionary Theory

Elements in the Philosophy of Biology

DOI: 10.1017/9781108591508
First published online: February 2020

Robert N. Brandon and Daniel W. McShea
Duke University

Abstract: In this Element, we extend our earlier treatment of biology's first law. The law says that in any evolutionary system in which there is variation and heredity, there is a tendency for diversity and complexity to increase. The law plays the same role in biology that Newton's first law plays in physics, explaining what biological systems are expected to do when no forces act, in other words, what happens when nothing happens. Here we offer a deeper explanation of certain features of the law, develop a quantitative version of it, and explore its consequences for our understanding of diversity and complexity.

Keywords: Evolution, random walks, null model, diversity, complexity

ISBNs: 9781108716680 (PB) 9781108591508 (OC)
ISSNs: 2515-1126 (online) 2515-1118 (print)

Contents

1 Introduction

Adaptation, diversity, and complexity – the three signature concepts of evolution, the three drivers of evolutionary research, the three great mysteries of life on Earth. Everyone knows they are distinct, but we are prone to blurring the distinctions, to assuming a close association among them, empirically and even conceptually. If something is diverse or complex, it must also be adapted, and vice versa.

The empirical separation is obvious and easy to demonstrate. Hugely expanded forebrains are adapted and complex, but the hominids that have them are not especially diverse. Red blood cells are diverse among vertebrates and well adapted for carrying oxygen, but these cells are not complex. The dinosaurs were diverse, with anatomy and physiology as complex as any animal's, but in the end they were not very well adapted (at least, not adapted enough to survive the late Cretaceous bolide impact).

Less obviously, there are conceptual separations that we often need to be reminded of, between adaptation on the one hand and diversity and complexity on the other. First, they are not just distinct; they are incommensurate. Adaptation is a relationship between an organism and its environment. An adapted organism is one that is likely to produce many offspring in that environment. Diversity and complexity are not relationships of any kind. They are absolute measures. They are numbers. Diversity is number of taxa (or, as we discuss later, degree of differentiation among taxa). Complexity is number of part types (or degree of differentiation among parts).

Second, adaptation on the one hand and diversity and complexity on the other are not even concepts of the same order. Adaptation is what one might call a first-order concept, like size, with each entity having some value attached to it. In contrast, diversity and complexity are second-order concepts. They apply only to groups or sets of entities. Diversity refers to differences within a group of organisms, species, or higher taxa. Complexity is differences among a set of parts within an organism. Diversity and complexity are variance concepts.

For adaptation, the body of theory that has been built up since Darwin is huge and robust. And nothing we have to say in this Element challenges a single principle or core claim of that body of theory, at least none that have been articulated. Our treatment is Darwinian, through and through. On the other hand, we do challenge one of the reflexes, the tics, of contemporary Darwinism, the very un-Darwinian tendency to assume that natural selection, and natural selection alone, governs all of evolution. Stephen Jay Gould and Richard Lewontin's (1979) famous Spandrels paper savaging this tendency had a salutary effect for evolutionary thinking in general, but it was less effective for diversity and complexity in particular. For these two, the common reflex assumption is still that selection rules. In fact, as we have shown in our earlier book and hope to amplify in this Element, selection is one of two important factors governing diversity and complexity. The other is the zero-force evolutionary law (ZFEL). The ZFEL is not inconsistent with Darwinian theory. Indeed, as we will show, the underlying principle is implicit in that theory. But theory routinely ignores it. The error, if we choose to call it that, lies in the application of Darwinian theory to diversity and complexity, seeing adaptation as sufficient when it is not. Thus, for contemporary Darwinian theory and its treatment of diversity and complexity, our summary statement is this: good theory, bad application.

We need to say a few words about the title, about the phrase "missing two-thirds." Evolutionary thought has a general theory of adaptation, a principle – natural selection – that explains adaptation wherever and whenever it occurs. Our claim is that what is "missing" is a general theory of diversity and complexity. Now there has certainly been a great deal of empirical work on diversity, at all scales, from the very recent and rapid human-driven global decline to the comparatively slow and fitful rise of diversity over the history of life. There has been much less but still *some* empirical work on the trajectory of complexity. And there has been much theorizing about the factors at work in particular historical episodes of change in diversity, such as the Cambrian explosion and the Permian and late Cretaceous mass extinctions. Again, there has been less for complexity, although the rise of metazoan complexity in the Cambrian has attracted some attention. But there has been no general theory, no set of principles laid out, that either explains or lays the foundation for explaining *all* change in diversity and complexity, wherever and whenever it occurs, nothing like what we have for adaptation. Our first book, in which we offered a qualitative formulation of the ZFEL, was an attempt to develop such a theory. In this Element, we extend that treatment a step further, offering a theory of diversity and complexity, one that is not only general but quantitative.

What about the "two-thirds"? Having said that there are three concepts, we may now have to backtrack a bit. In standard usage, *diversity* and *complexity* are different. *Complexity* is within-organism variance. *Diversity* is among-individual or among-taxon variance. And from that perspective, they make up two-thirds of the grand trio – adaptation, diversity, and complexity. But one could also see diversity and complexity as the same concept, variance, applied at different levels of the biological hierarchy. In that case, it would be more accurate to say that only half of evolutionary theory is missing (which in a way would be a relief, making the void seem less frightening). But whether two-thirds or a half, the gap strikes us as serious. If modern theory has not been blinded by its selectionist impulses, it has certainly been blinkered by them, glimpsing the factors underlying diversity and complexity only incompletely and fleetingly. Here, as in our first book, our hope is to bring them forward, to position them in the center of theory's visual field, where they cannot help but be noticed.

Section 2 presents what we call the zero-force evolutionary law. In this section we give a qualitative version of the law, which is what we developed in our earlier book (McShea and Brandon 2010). Just what it means and why it is true are explained. Having said what the ZFEL is, in Section 3 we briefly say what it is not. In particular, we cover what we think have been the most prevalent misunderstandings of our earlier work. Section 4 gives two different, but closely related, quantifications of the ZFEL. Quantification of the ZFEL has two main virtues. First, it offers a clear mathematical demonstration of the basic claim underlying the ZFEL, namely, that two independently evolving entities will tend to diverge. This claim is counterintuitive (since two random walks should converge as often as they diverge), but our mathematical models prove it to be true and, furthermore, help educate our faulty intuitions about pairs of random walks. Second, a quantified ZFEL is able to play a special role in a zero-force theoretical framework, analogous to the role of Newton's first law in physics. If we can quantify just how much two lineages should diverge, then we can say with some quantitative precision what forces need to be invoked when observations fall outside of ZFEL-based expectations.

That is all abstract theory. The implications of this for biology are discussed in Section 5. Having developed the theory necessary for testing the role of selection in the evolution of diversity and complexity, we could stop and wait for the evidence from numerous studies of particular cases of diversity or complexity to build up. That is the epistemologically cautious thing to do, and we do approve of that attitude. But we also think that we know enough to make some educated guesses. We come to somewhat different conclusions with respect to diversity on the one hand and complexity on the other. The ZFEL has certainly pushed complexity forward in evolution, but we cannot account for the actual trajectories of the evolution of complexity without invoking pretty consistent selection against complexity. In contrast, there are good reasons to think that the diversity of life at the largest scale – globally, on timescales of hundreds of millions of years – is governed *mostly* by the ZFEL. The claim arises from a combination of the theory and some simple facts about biology. It is an empirical claim, one that can in principle be checked using the quantitative formulation. This pair of conclusions may seem odd, since we have argued that diversity and complexity are one and the same thing but viewed from adjacent hierarchical levels. We show that, however odd, it is consistent and, beyond that, quite likely.

2 The Zero-Force Evolutionary Law

2.1 Zero-Force Laws

Aristotle and Newton looked at the physical world in radically different ways. And the word *radical* in the previous sentence almost certainly understates the case. It is hard for scientifically literate people living in the twenty-first century to fully appreciate the fundamental differences in conceptual frameworks used by Aristotle and Newton. For Aristotle, the natural state of physical objects is rest. Motion (whether accelerative or not) requires the imposition of a force. In contrast, for Newton, the natural state of physical objects is constant velocity (with $v = 0$, i.e., rest, counting as constant velocity). Living on the face of Earth, with its viscous atmosphere, it is hard to see who is right. Sitting in an armchair, as philosophers are wont to do, every visible thing is at rest. This fact fits Newtonian and Aristotelian physics equally well. (For Aristotle, these objects – pen, notepad, laptop, etc. – are in their natural state; for Newton, they are moving with constant velocity equal to 0.) Looking out the window, where right now there happens to be a hurricane blowing, leaves, branches, and the occasional small animal are flying past. Aristotle would claim, and he would be right, that these flying objects depend on a constant force being imposed upon them. Of course, Newton agrees, because he would see this case as one where the viscosity of the air produced a backward directed force on the objects that the forward directed force of the wind must overcome.

We want to draw two points from this. First, settling on the correct view of the natural state of motion, or the correct zero-force law, is not a straightforward empirical issue. Newton's first law is not a statistical generalization. (See Nabi 1981 for a hilarious spoof of what physics would look like if the laws of motion were decided by statistical methods.) Rather, the issue between Aristotle and Newton is theoretical – which theoretical framework works best in doing physics? With the benefit of hindsight, we can say that the move to the Newtonian framework was a great advance for physics. The second point has to do with the advantages of a quantitative theoretical framework. Aristotle had nothing close to a quantitative theory of physics. Newton's physics was quantitative, so that he could, for instance, go beyond the qualitative statement that the force of aerodynamic drag works in the opposite direction of the force of the wind operating on the squirrel that just blew past. These forces can be assigned quantities so that we can make quantitative predictions about direction, velocity, and acceleration of the squirrel.

In particular, when we know the quantitative value of one force, we can deduce the quantity of the second opposed force, but only in the context of a zero-force law. For instance, near the surface of the Earth, we know the quantity of the gravitational force acting on a falling object, and so we can deduce the force of drag from Newton's first and second laws (the second law being $F = ma$). More generally, any quantitatively characterized deviation from inertial motion (described in the first law) tells us the quantity of the net force acting on the system. This is our goal for the ZFEL: describe the zero-force condition for diversity and complexity and then use that in conjunction with observations to deduce net forces, such a natural selection. Immodestly put, we aim to do for biology what Newton did for physics.

One important consequence of this needs emphasis. In statistical sciences, there is an important distinction between two methodologies. One is null-model testing, where one sees whether observations are, or are not, significantly different from what the null model predicts. There are only two possible conclusions from such a test: either we reject the null (but are given nothing in its stead) or we do not reject the null (which, of course, does not mean we should accept the null). If one's goal is to find some hypothesis that is reasonable to believe, then this methodology is not adequate. A second methodology can deliver reasonable beliefs, namely, the methodology of maximum likelihood testing. Here we compare two or more hypotheses to the data. Each hypothesis has a likelihood given the data, and this methodology tells us to accept the one that has the highest likelihood. Null-model testing is great for those wishing to avoid false beliefs, and maximum likelihood is great for those wishing to maximize their true beliefs. There is no accounting for matters of epistemological taste.

Null-model testing within the context of a zero-force theory has all the epistemological virtues of null-model testing more generally, as discussed above. But it also delivers an alternative hypothesis when the null is rejected. Consider Newtonian physics. When a massive object is observed to be accelerating, we reject the Newtonian null of no acceleration, but we are then immediately given a (quantitative) hypothesis of the net force acting on the object. It is true that we are not given the nature of this net force – we are not told whether it is gravity or electro-magnetism or something else – but we are given the direction and magnitude of this net force. A zero-force theoretical structure gives us a way to have our epistemological cake while eating it. One might be disappointed in the cake, that is, one might be disappointed that the hypothesis delivered does not specify the nature of the force, but even this is oftentimes mitigated. For instance, in the physics of planet-sized objects, we can usually hone in on gravity. And in biology, we can often implicate selection. Here is the short version of this Element: don't take selection as your default hypothesis; only invoke selection when stochastic behavior of independently evolving entities (the ZFEL) has already been factored in and there is some remainder left to be explained.

2.2 The Principle Underlying the ZFEL

The ZFEL says that in the absence of imposed forces and constraints, diversity and complexity tend spontaneously to increase. The reason is that variation accumulates, with the result that entities tend to become different from each other, to diverge. When these entities are organisms or taxa, this divergence is an increase in variance and therefore an increase in diversity. When the entities are parts of organisms, the divergence is an increase in complexity.

The principle is simple and familiar to everyone, at least in nonbiological contexts. A group of cars that are essentially identical the moment they come off the assembly line will be treated differently, suffer different accidents, and, as a result, look very different from each other years later. They diverge. The variance in form among them increases. A group of kindergarteners released onto a playground at recess time tends to spread out, to disperse throughout the playground with time. The variance in their locations increases. Daughter languages diverge from the parent language and from each other over time. The variance among them in spelling, usage, idioms – in every feature of a language – tends to rise. They diverge.

The principle is familiar, as we say, but its application to the structure of organisms (complexity) and the structure of taxa (diversity) is not straightforward. In this section, we try to explain exactly how it can be done.

2.3 The ZFEL

Here is the formal statement of ZFEL, from McShea and Brandon (2010):

> ZFEL (special formulation): In any evolutionary system in which there is variation and heredity, in the absence of natural selection, other forces, or constraints acting on diversity or complexity, diversity and complexity will increase on average.

We called it a "special formulation," because it applies in, but not only in, the special case where forces and constraints are entirely absent. Notice that the phrasing parallels Newton's first law, describing the behavior of objects in the absence of forces. The ZFEL says that in the absence of forces, diversity and complexity increase, on average. Newton's first law says that in the absence of forces, objects travel with constant velocity. Both this formulation of the ZFEL and Newton's first law are in-principle statements, designed to tell us what happens when nothing happens (i.e., when no forces impinge on the system).

But we hasten to emphasize the fact that the special formulation of the ZFEL does indeed apply to cases beyond the ideal case of no forces and no constraints. That is good, because as we point out below, the case of no constraints is a conceptual impossibility. In contrast, the case of no selection is conceptually possible, though perhaps empirically rare. However, the special formulation above does not have such a limited application. Notice it describes the absence of forces or constraints acting on diversity or complexity. Consider an example of two lineages, A and B, and a trait, mean height, that differs between them. (This is a diversity example, for a complexity example take A and B to be two parts of a common whole, e.g., two vertebrae in a mouse.) The special formulation of the ZFEL applies when selection acts on some other uncorrelated trait, say, color. In that case, a force is present, but it is not acting on the trait in question, height. The special formulation also applies when selection acts directionally on height in both lineages, provided it does so in each lineage independently, that is, provided it is not directly favoring the two lineages becoming more similar or directly favoring their become more different. Thus when selection acts independently on each lineage, a force is present, but it is not acting directly on diversity. In sum, the special formulation *does not* apply to cases where selection acts directly on diversity. That is what the absence-of-force clause in the special formulation rules out.

Thus, whether the special formulation of the ZFEL has a wide or narrow range of application is an empirical matter, one that cannot be settled by the bromide that selection and constraints are always present.

So to cover a broader range of cases, we also gave a general formulation:

> ZFEL (general formulation): In any evolutionary system in which there is variation and heredity, there is a tendency for diversity and complexity to increase, one that is always present but may be opposed or augmented by natural selection, other forces, or constraints acting on diversity or complexity.

The critical word in this formulation is "tendency." When forces or constraints are present, diversity and complexity have a tendency to increase. A tendency is a kind of push or straining toward increase. It is not an actual outcome. In other words, the general formulation of the ZFEL does not say that diversity and complexity *will* increase, only that there is a kind of pressure, or oomph, toward increase. Analogously, if I lean against my house, I impart to the house a tendency to fall down. The house does not fall down, indeed it does not even come close to falling down, because my leaning is resisted by various forces and constraints, but the tendency is there nonetheless, so long as I am leaning against it. In the same way, the ZFEL says that lineages and parts have a tendency to diverge from each other, even if – owing to selection and constraints – they do not diverge. The implication of course is that the moment those forces and constraints are removed, divergence would begin.

Our usage of the word tendency is conventional, but we note that occasionally, in some contexts, the word is also used to describe an actual trend. One hears it said that the stock market has shown an upward tendency over the past year, meaning simply that some indicator has increased, that there has been a trend In our understanding, however, the fact of the trend by itself gives no information about a tendency. There could have been an increasing tendency, imparted perhaps by market forces, but the trend could also have been due entirely to chance, with no underlying upward tendency. In any case, here, for the ZFEL, a tendency is a kind of predisposition to increase, not a result, not an actual trend. We say more about this shortly.

2.4 A Simple Model

The model in Figure 1A shows why the ZFEL works. When forces and constraints are absent, evolving entities – lineages (for diversity) and parts (for complexity) – change randomly, and as a result they diverge from each other, on average. The figure shows a group of 10 entities, changing in a size morphospace. As shown, they all start at the same size, 20 mm. For complexity, the figure might represent 10 teeth in a reptile's tooth row, all initially 20 mm long. For diversity, it might represent body length in a population of 10 individuals of some insect species, all 20 mm long.

In the model, each entity changes randomly, increasing or decreasing in length by 1mm in each time step, always with 50:50 probability. The result is 10 random walks, shown over 30 time steps. Notice that at the end of 30 time steps, the random walks have dispersed considerably. The histogram above the trajectories shows what the expected distribution would be if there had been hundreds of random walks, rather than just 10. The Central Limit Theorem tells us that the histogram approximates a normal distribution, with variance equal to the number of time steps. Thus diversity or complexity – here measured as the variance –increases without limit. In other words, the ZFEL expectation is not only that diversity and complexity increase initially, when lineages or parts are quite similar to each other, but also later, when they have become quite different. Even for a diverse set of individuals or species, the ZFEL expectation – in the absence of forces or constraints – is even greater diversity. And for individuals that are already quite complex, having highly differentiated parts, the expectation is even greater complexity. Of course, the expectation is probabilistic. In any time step, diversity or complexity may decrease, if by chance

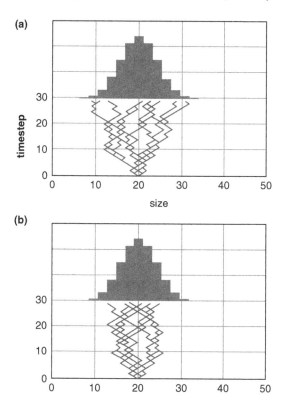

Figure 1 (A) Special formulation of the ZFEL. Ten entities following independent random walks over 30 time steps. The histogram shows the distribution of outcomes for a much larger number of random walks. (B) General formulation of the ZFEL. Trajectories of the 10 entities are limited by selection and constraints (see text), but the *tendency* to spread is present the entire time. (The histogram from A – the expectation due to the ZFEL – is copied here to enable a comparison of the actual with the expected.)

entities vary in such a way as to become more similar to each other. But the expectation, the on-average result, is always increase.

No absolute timescale needs to be specified here. What we call *time steps* can be thought of as generations, or million-year increments, or even longer units of time. The ZFEL principle operates on all timescales. Also, here and in what follows, the horizontal axis is treated as an additive scale, but recognizing that changes in biology tend to be proportional, one could instead interpret it as a log scale, with increases and decreases occurring in units of constant proportion.

Figure 1A illustrates the special formulation of the ZFEL, showing divergence when selection and constraint are absent. Figure 1B shows one way that the general formulation of the ZFEL could become relevant. Suppose the evolving entities are 10 teeth in a tooth row and that selection opposes any decrease in tooth size below about 15mm, perhaps because a smaller tooth is unable to apply enough force to properly pierce an important food, say, a hard seed. Also suppose that a developmental constraint imposed by jaw size limits tooth size to a maximum of about 25mm. In this version of the model, teeth still change randomly, and divergence still occurs, driving complexity initially upward, but divergence is eventually limited by selection below and constraint above. As a result, the divergence histogram from Figure 1A, reproduced in Figure 1B, is not filled out by time step 30. In other words, divergence at time step 30 – again as measured by the variance – has been less than the ZFEL would have predicted in the absence of selection and constraint.

Figure 1 provides an opportunity to illustrate what we mean by a tendency. The general formulation of the ZFEL says that the *tendency* for tooth sizes to diverge is present over the entire 30 time step span. Initially the tendency is manifest, producing some divergence and an upward trend in complexity (over about the first 10 time steps). Our point is that it is still present later, even when invisible, when actual divergence is blocked (over the last 20 time steps). An implication is that if either limiting factor were removed in the thirty-first time step – say, if selection against small teeth were removed by a change in diet – divergence and the trend in complexity would resume.

Figure 1 also gives us a chance to preview of how the ZFEL can used to detect and measure imposed forces and constraints. In Figure 1A, the complexity of the set of teeth at the end of 30 time steps is the variance of the terminal histogram, and that variance is the ZFEL expectation. In other words, it is the expected complexity in the absence of selection or constraint. But if the actual divergence of the teeth follows a pattern like Figure 1B, producing less variance after 30 time steps than predicted by the ZFEL, then that difference in variance – assuming it is statistically significant – can only be the result of selection or constraint. What is more, the magnitude of the difference between the ZFEL-predicted variance and the actual variance is a measure of the intensity of selection and constraint, the intensity with which complexity has been opposed. And we can apply the same principle to detect and measure forces and constraints in cases where the actual variance is *greater* than the expected, where selection or constraint have favored diversity or complexity. We explain further in Section 4.

2.5 Constraints: Constitutive, Imposed, and Effective

The notion of "absence of constraint" in the special formulation of the ZFEL may be puzzling. A property of an object is a kind of constraint. A bear that is white now is constrained not to be simultaneously brown, or any other color, simply as a matter of logic. So, an object that is literally without constraints would be one without properties, which is unimaginable. Furthermore, in evolution, the total absence of constraint would mean the absence of heredity. And in the absence of heredity, a white bear could give birth to a bear of any color whatsoever. Indeed, it could give birth to a crocodile or a toaster oven. The point is that the intent in the special formulation is not to exclude what might be called constitutive constraints, constraints that are essential to biology and to our ability to imagine and understand the system, such as logic and heredity.

Rather, it is to exclude what we call imposed constraints. These are constraints that block the accumulation of variation. For example, in marine mollusks, there is a constraint on diversity imposed by planktotrophic feeding, a life habit in which larvae disperse widely, discouraging the isolation of populations that disposes a taxon to generating new species. In the language of the ZFEL, planktotrophic feeding is an imposed constraint that limits diversity. It is constraints like these that must be absent for the special formulation of the ZFEL to apply. Imposed constraints take many different forms, including the ones commonly recognized in biology – e.g., developmental, phylogenetic, architectural – but also physical and mathematical. A phylogenetic constraint limits the number of body regions in insects to three. And because these three are already differentiated in all species (into head, thorax, and abdomen), the phylogenetic constraint leads to a mathematical constraint on the body-region complexity of insects. A system simply cannot have more part types than parts. Notice that imposed constraints can always be broken or avoided, at least in principle. The constraint on diversity imposed by planktotrophic larvae can in principle be broken by switching to a nonplanktotrophic life habit. The mathematical constraint on insect body-region complexity can be avoided by the addition of a fourth body region.

Of course, not every imposed constraint on diversity and complexity will actually be limiting. A phylogenetic constraint limits the number of neck vertebrae in most mammals – from mice to giraffes – to seven. (The giraffe ones are really *big*.) And this limit is also a mathematical constraint on the skeletal complexity of the neck in that it limits the number part types to seven. However, the standard categorizations of mammalian neck vertebrae recognize only three types. The first vertebra, the atlas, is modified for articulation with the skull, and the second, the axis, modified for articulation with the atlas and to allow greater rotation of the neck (compared to reptiles, for example). The remaining five vertebrae are a single type, all very similar to each other. So, the imposed constraint limits number of part types to seven, in principle, but in fact only three types are presently realized, and therefore the imposed constraint is not truly limiting. A fourth vertebral type could arise without adding more vertebrae. In our terms, we say that the constraint is imposed but not effective. Owing to the huge redundancy present in organisms, many imposed constraints are not effective. For example, an imposed mathematical constraint limits protein diversity. Proteins consisting of some number of amino acids, say, L, are limited in their diversity to L^{20}, where 20 is the number of standardly available amino acid types. But in no group of organisms is this theoretical limit on diversity even approached, with the result that this imposed mathematical constraint is not effective. A drunk walking down the middle of a narrow alley will eventually hit the gutter. A drunk walking in the middle of Tiananmen Square may well be arrested, but will not end up in the gutter. Only effective constraints actually limit diversity and complexity.

In sum, only a subset of all possible constraints are targeted by the constraint-exclusion clause of the special formulation of the ZFEL, namely, those that are both imposed and effective and not constitutive of the system.

2.6 Randomness

The divergence among entities predicted by the ZFEL follows from that the fact that they change randomly when forces and constraints are absent. Here we say a few words about how we understand randomness. (For a longer discussion, see McShea and Brandon 2010.) In population genetics, random change ordinarily means drift. And in fact, standard thinking in biology predicts that drift leads to increased diversity (between, not within populations). But for the ZFEL, the relevant notion of randomness is broader than drift. It includes the case in which entities at any scale change deterministically but independently of each other, that is, in which they change *randomly with respect to each other*. (We alluded to this earlier, in the discussion of the domain of application of the special formulation, and here we explain further.)

Consider a nonbiological case. A group of teenagers wear the same uniform to school every day but are told that on one special day that they can wear whatever they like. On that day, each student makes his or her own choice of clothing deterministically, each following his or her own preferences (perhaps, one would hope, limited somewhat by their parents). The result is a great diversity of outfits on display at the school. The students are choosing deterministically but randomly with respect to each other – that is, independently of each other – and the consequence is a rise in diversity. At this point, a reasonable question would be: if this is randomness for purposes of the ZFEL, what is the contrast case? What is nonrandomness? For the ZFEL, the nonrandom case would be one in which diversity rises because some pervasive force is acting, some force that acts directly on diversity. In the case of the students, this might happen if groups of them talk among themselves ahead of time and agree on a choice of outfit. In that case, diversity might increase but be less than it would be without collusion. It might also happen if students *try* to dress uniquely, if each student picks his or her outfit in an effort to look outlandish, to look as different from the other students as possible. In that frightening to contemplate case, diversity would increase more than if their choices were simply independent. In other words, nonrandomness requires some sort of causal linkage among individuals, here teenagers, but for the ZFEL the entities that form lineages.

Turning to biology, consider two closely related snail species, one changing under selection for a wider aperture and the other under selection for a thicker shell. In other words, they are changing randomly with respect to each other. Both are under deterministic selection, but independently, which means they become more different from each other but without any force acting directly on their differences. There is selection, and diversity increases, but there is no selection *for* diversity. Thus the zero-force condition of the special formulation of the ZFEL is met, and the resulting increase in difference between them is attributable to the ZFEL. Here a contrast case would be two species competing for the same resource and simultaneously selected for reduced resource overlap, for avoiding competition. In that case, selection is directly *for* their differentiation, the zero-force condition is not met, and selection – not the ZFEL – is the cause of the divergence. Both are recognized as important mechanisms for producing divergence at the species level, both the ZFEL – that is, independent selective forces acting on different species – and competition-driven divergence. However, in Section 5, we argue that to explain the bulk of macroevolutionary diversity, the ZFEL is the only possible mechanism.

Randomness in the with-respect-to sense is not mysterious in any way, nor is it foreign to biology. Indeed, drift itself can be understood in this way. Consider the case in which an allele that is not under selection decreases in frequency in a small population. Zooming in on the process, we might see that one individual with the allele was swept away in a flashflood. Another failed to find a mate – for reasons causally unconnected with the allele – and did not reproduce. Yet another produced offspring but they were lost to an unseasonal cold snap. In other words, there were deterministic causes acting in each fitness failure, but the causes were different and independent of each other, and none related to the relevant allele. They are random with respect to each other. And that is drift. Let us more to the molecular level. When the vicissitudes of meiosis produce in some small population a disproportionate representation of some allele, the possibility always exists that causation is truly random, in other words, that drift is the result of quantum effects percolating up to the macromolecular level. But quantum effects need not be invoked for drift to occur. It could be that if we look closely enough, deterministic physical causes would be found in every meiotic event. In that case, calling the result drift would not be a claim that no forces were acting, only that no concerted, directional forces were acting, no forces favoring one allele over the other. In sum, where quantum percolation is not involved, drift as conventionally understood is multiple causes acting independently. It is randomness in the with-respect-to sense.

2.7 Diversity, Complexity, and Adaptation

The ZFEL is a law about diversity and complexity. It has consequences for adaptation, which we explore later, but it is not about adaptation. And therefore the understandings of diversity and complexity on which it is based need to be adaptation-free. Happily, in standard usage, diversity is already an adaptation-free concept. We call a group of organisms diverse if it has many species, regardless of how well adapted they are. The great diversity of brachiopods in the Paleozoic is diminished not at all by the fact that they were about to be replaced by the (arguably) better adapted clams. And conversely, the low diversity of clams throughout the Paleozoic says nothing at all about their later-to-be-demonstrated fitness. Diversity is just a count, of number of kinds of individual in a population, of number of species in a genus, or more generally, number of taxa in some higher taxon.

This is not to say that adaptation is irrelevant to diversity. Indeed, some of the central questions of evolutionary studies have to do with the effect of various environmental regimes on diversity, the consequences of diversity for the occupation of adaptive zones, and the capacity of organisms with various adaptations to diversify. But these connections are causal, not definitional. And thankfully so, because definitional entanglement would obstruct the investigation of causes. Suppose that our measure of diversity were weighted somehow by an assessment of fitness, that the diversity of late-Paleozoic brachiopods were measured not as a simple count of species but as a count that is downweighted by some fitness factor corresponding to their apparent inferiority to clams. And now suppose we wanted to ask about the causal relationship between the greater adaptedness of clams and its effect on the "diversity" – in this strange sense – of brachiopods. Did clams displace brachiopods on account of their superior adaptedness? The circularity is obvious and unavoidable. We would be asking, in part, about the relationship between a variable and itself, between adaptedness and "diversity," with "diversity" measured now in a way that includes a component of adaptedness.

The point may seem obvious, even silly. Who would think to deliberately introduce a definitional entanglement between diversity and adaptation? We

♦

have bothered to make the point, because the colloquial understanding of complexity bizarrely creates exactly this entanglement! Complexity in its colloquial usage outside of biology – and until recently, within biology as well – has a component of adaptedness built into it. We call human beings "complex" not only because of some aspect of our structure, such as our large number of cell types, but because we imagine ourselves to be impressively capable, exquisitely functional, that is, well adapted. Now suppose that we wanted to ask whether our complexity in this sense is the result of natural selection, of a process of adaptation? Using the colloquial understanding of complexity, we will never know. In an empirical study, every rise in "complexity" in our evolutionary history will be discovered to be at least partly the result of adaptation, and this would be so *by definition*.

What to do? The sensible thing to do is to disentangle complexity from adaptedness, to centrifuge these immiscible concepts to purity. We need a concept of complexity that is distinct from the colloquial one, one that is a measure purely of structure, with no contribution at all from adaptedness. A number of alternatives are available. The one we have adopted here is to measure complexity as number of part types. This approach has proved useful and has been adopted in biology, mostly in the investigation of trends in complexity. For example, Valentine et al. (1994) used number of cell types to document a trend in complexity in metazoans, and Sidor (2001) used number of skull bone types to document a (downward) trend in the complexity of tetrapod skulls. It is now also a standard usage in molecular biology, used to describe changes in numbers of genes or numbers of protein types involved in various molecular mechanisms (Doolittle 2012; Finnegan et al. 2012).

So complexity in the sense of part types is a purely structural concept, uncontaminated by any notion of function or adaptedness. Staying true to this definition has some odd-sounding consequences. A dead organism could be more complex than a live one of the same species, say, if the decay process after death produces some further differentiation among cells, and therefore more cell types. (In the decay process, cells will eventually disappear, of course, constraining complexity and ultimately reducing it to zero, but an increase in complexity is plausible in the short run.) However upsetting this consequence – death producing an increase in complexity – we can console ourselves with the knowledge that the conceptual surgery that replaces colloquial complexity with pure complexity in our scientific vocabulary is only painful in the short run, and that the gains will be worth the suffering. With pure complexity we have at our disposal an operational metric, a tool, for investigating trends in complexity, correlates of complexity, and causal relationships between complexity and almost any variable we are interested in, including adaptation. In any case, are the consequences really so disturbing? A dead organism can be warmer or more massive than the live one, so why not greater in other physical measures, like complexity. Furthermore, colloquial complexity was never a scientific concept anyway. It has something to do with structure and degree of adaptedness, but no one has ever devised a clear statement of what it is or a way to measure it in real organisms. Finally, as almost everyone has long known, at some level of awareness, colloquial complexity has always been a charade, a word that gives a scientific flavor to the long discredited Great Chain of Being. It is a way to sound scientific while bragging about human superiority, without having to bother explaining precisely what that superiority consists in.

For our purposes, complexity in the sense of part types is sufficient. But it is worth noting that there are other purely structural, adaptation-free ways to understand complexity. Instead of part types, complexity could be a count of number of independent processes, or of independent interactions, occurring within an organism (McShea 1996). There is also complexity of development, the number of independent processes involved in generating an organism. We do not explore these further here, but clearly the ZFEL would also apply to complexity in these senses. Finally, there is a complexity in a very different sense, number of levels of nestedness in an organism, of parts within wholes, also called levels of organization. A multicellular individual is one hierarchical level deeper, one level of complexity higher, than a single-celled individual. A colony or a society is two levels higher. We will have more to say in Section 5 about the relationship between complexity in this sense and complexity in the sense of part types. For now, we just need to point out that the ZFEL does not apply in any straightforward way to hierarchy. That is, it makes no prediction about whether number of levels will increase, or decrease for that matter, over time. Indeed, for our purposes, the fact that the word complexity has this double usage, for both part types and levels, is an unfortunate source of confusion. To minimize the risk, various tactics are available. One is to use the phrase "vertical complexity" for number of levels and "horizontal complexity" for number of part types within a level (Sterelny 1999). Another – and the one we adopt here – is to use the word complexity – alone and unmodified – for the horizontal part-type sense and to avoid the word complexity entirely for the vertical sense, using instead the words level or hierarchy.

Before closing this section, we want to acknowledge that the word "part" is not generally recognized as a technical term in biology. Of course, the notion that organisms can be decomposed into parts would be relatively uncontroversial for parts at some scales. For whole organisms, cells are parts, for example. At other scales, there is more ambiguity. In a human, is an arm a part, and if it is, are the muscles that connect it to the shoulder also parts? Worse, are those muscles to be included in the *same* part with the rest of the arm? Answering these questions may seem hopeless but in fact a graph-theoretical definition of a "part" has been devised, based on degree of connectedness among elements, and protocols for identifying parts in difficult cases have been developed (e.g., McShea and Venit 2001).

2.8 Discrete and Continuous Measures

So far, both diversity and complexity have been treated in their discrete senses. Diversity is number of different types of individual in a population or of species or taxa in some higher taxon. Complexity is number of part types within an individual. But both concepts also have a more general sense, the continuous sense.

For diversity, the continuous measure is degree of differentiation among entities, or what in paleobiology is called *disparity* (Foote 1994). The mammalian order Perissodactyla is disparate in that horses, tapirs, and rhinoceroses are phenotypically quite different from each other, but not discrete-sense diverse, because the group has few species. The modern Pholidota, the pangolins, are neither disparate (because all are very similar to each other) nor discrete-sense diverse (only eight extant species). These comparisons are impressionistic, but a number of quantitative measures have been used in disparity studies (e.g., Ciampaglio et al. 2001). One is the standard deviation or variance among entities in some measure of the phenotype, such as length or body mass. Other metrics, such as average absolute difference and range of variation, have also been used. Typically, disparity is measured over a large number of dimensions and comparisons of disparity are done in a multidimensional morphospace. Disparity is a more general measure of diversity than counts of entity types, because it can be used whether or not entities are cleanly sortable into discrete bins.

Like diversity, complexity has a more generally applicable continuous sense, the degree of differentiation among parts. The complexity of a tooth row is the degree of differentiation among the teeth in it. Again, like diversity, it can be measured using standard deviation or any of a number of variance analogues, and it can be applied to parts varying in one dimension, or in a number of dimensions so that comparisons can be done in a multidimensional morphospace. McShea (1993) used variance analogues, including absolute difference and range of variation, to investigate evolutionary change in the complexity of mammalian vertebral columns.

At the outset we said that both diversity and complexity are variance concepts, that they are second-order concepts, and this distinguishes them in an important way from first-order concepts like adaptedness and body size. Here we add that both diversity and complexity can be measured with the same ruler, the statistical variance (or with any of a large number of other variance measures). And the reason is that they are precisely the same concept, applied at different levels of organization. In standard usage, diversity is variation among individuals and among taxa, while complexity is variation within individuals. It is merely an accident of language – and perhaps of human perceptual tendencies – that we have different words for the two. In other words, complexity just is diversity. Putting the two together in the same sentence, we might say that "the complexity of an organism just is the diversity of its parts." And, while not idiomatically acceptable, it would not be incorrect to speak of the complexity of a genus, understood as the diversity of the species within it.

2.9 Level Relativity

One more conceptual point needs to be made, one that is by now obvious perhaps. Both diversity and complexity are level-relative concepts, meaning that the diversity or complexity of a system at some level has no necessary relationship to its diversity or complexity at any other level. The complexity of a fish at the level of tissues and organs is about 90, meaning that it has about 90 different tissue and organ types. Its complexity at the level of cell types is about 130. And at the atomic level it is 6, if we include just the most abundant atom types: oxygen, carbon, hydrogen, nitrogen, calcium and phosphorus. The point is that a fish has a different complexity at every level, and that these numbers are in principle independent of each other. An animal could in principle have thousands of cell types and only a few tissue types. Or vice versa, it could have thousands of tissue types and a much smaller number of cell types, if each tissue were composed, say, of a unique combination of those few cell types. Furthermore, change in complexity is independent across levels, at least in principle. Complexity at the molecular level (number of molecule types) could go down while number of cell types goes up, and vice versa. In fact, as we shall see in Section 5, an inverse relationship of this type has been demonstrated: the cells of complex multicellular organisms are simpler – have fewer internal part types on average – than free-living single-celled protists. All of this is true of diversity as well. When the diversity of some group measured in terms of genera goes up, the diversity of species in the same group could either go up or down.

Finally, on account of level relativity, there is no preferred level of analysis. A genus has no true diversity. The number of species it contains is no more its "real" diversity than the number of types of individual it contains. And likewise, an organism has no true complexity. In particular, its complexity at the genetic or molecular level is no more its real complexity than number of cell types or its complexity measured at any other level.

2.10 Other Matters Arising

Before turning to the quantitative formulation, a number of lesser issues need some mention. All are treated at greater length in McShea and Brandon (2010).

Universality. The ZFEL for diversity applies to all populations and taxa, in all places on Earth and at all times in the history of life. The ZFEL for complexity applies to all characters, in all organisms, everywhere and always. More generally, it applies to all evolutionary systems, all systems in which there is variation and heredity. Notice that these two requirements are a subset of Darwin's three conditions for the occurrence of selection itself: variation, heredity, and nonrandom differential reproductive success. It follows that the ZFEL is at least as widely applicable as selection, while logically it is more basic.

Relationship to Standard Darwinian Theory. The principle underlying the ZFEL is not new to biology. In particular, the notion that randomly varying entities spontaneously diverge is implicit in standard models in molecular biology – such as models of the divergence of duplicate genes – and in phylogenetic systematics – in random models of lineage divergence in character space. A ZFEL-like principle has been recognized in many other contexts as well, but it has not been formalized, nor have the connections among its various applications been explicitly acknowledged. Thus, what the ZFEL offers is a unification, and now in the present treatment, a quantitative unification. Furthermore, in Section 4, we will give a mathematical explanation of why divergence occurs, and, to our knowledge, this is new.

Natural Selection. We are realists with respect to causes. Causal processes and causal interactions are real, existing independently of our interest in or awareness of them. But as in our earlier book, here we adopt a conventionalist stance with respect to forces. One might think that forces are causes and that our realism with respect to causes commits us to a similar stance with respect of forces. But we take the term *force* to be well defined only within a theoretical framework. Newtonian mechanics offers us a paradigm of a theory of forces. In it, gravity is a force. But within the framework of General Relativity gravity is not a (straightforward) force. Within the theoretical framework of this Element, natural selection is a force.

3 What the ZFEL Is Not

Here we address three ways that the ZFEL could be, and has been, misunderstood. Others are addressed in McShea and Brandon (2010).

3.1 The ZFEL Makes No Empirical Claim

The focus of most publications in science is some empirical claim, some statement that the authors are asserting to be true about the world. And the heart of the publication is typically the explication of methods and the presentation and interpretation of data that support the empirical claim.

This Element is different. We do make empirical claims here, especially in Section 5, but the ZFEL by itself makes none. By this we mean that the ZFEL does not claim that modern organisms are more diverse or more complex than ancient ones. It does not claim that there has been a trend in diversity or complexity. Thus, evidence against the existence of a trend is not evidence against the ZFEL. It is not relevant to the question of whether the ZFEL is true.

We emphasize this point because the ZFEL has sometimes been misunderstood on this issue. For example, in an interesting and sometimes apt critique of our first book, Bromham (2011) worries that the existing body of biodiversity data "does not provide clear and unambiguous support for a tendency for diversity to increase over time, as predicted by the ZFEL." And then later she adds, "the ZFEL makes a prediction that most lineages should show a rightward-trend from simpler ancestors to more complex descendants. But this pattern is true in only a minority of cases." In conclusion, she writes, "But claiming that a tendency toward diversification is a universal law of biology weakens this book considerably ... By christening the tendency to divergence as a law, the authors are forced to jump through quite a lot of hoops to reject any cases where the tendency to diverge does not seem to be a general feature."

Remarks of this sort – not unique to Bromham by any means – misunderstand the kind of law we are proposing. The ZFEL is not a law in the sense of, say, Cope's rule, which claims that body size in general increases in evolution, or Coulomb's law, which says the repulsion between like charges follows an inverse-square function. Rather, it is a law like Newton's first law, which gives a framework for understanding motion. The law says that motion should be decomposed into a component of constant velocity and a component due to imposed forces. Huge numbers of accelerating objects do not falsify Newton's first law. Similarly, the ZFEL says that change in diversity and complexity should be decomposed into a component due randomness-driven divergence (the ZFEL) and a component due to imposed forces. It does not say that they will be exclusively randomness-driven in most – or even any – lineages.

To put this another way, Cope's rule and Coulomb's law would be called into doubt if new evidence revealed that body size does not increase in general and that the exponent governing repulsion between like charges was different from 2. And evidence that diversity and complexity do not increase in general would indeed be damning for the ZFEL if it did in fact predict a trend. But it does not. The ZFEL is a null model, one that tells us how diversity and complexity are expected to behave in the absence of forces. It makes no claim that forces are absent, and therefore no claim that a trend in either diversity or complexity has actually occurred. Indeed, the ZFEL would be perfectly consistent with a finding that diversity and complexity hardly ever increase in evolution, just as Newton's first law is perfectly consistent with a finding that objects hardly ever move with constant velocity. (Come to think of it, forces are never precisely zero, anywhere in this universe, so they *never* do.)

A related misunderstanding needs to be headed off. The ZFEL does not claim (cf. Deline et al. 2018) the evolution of organisms actually follows a random walk or that the evolution of any group is best approximated by a random-walk model like that in Figure 1A. Again, the ZFEL is a null model. It does not make any empirical claim about actual evolutionary trajectories. Rather, it gives a theoretical expectation to which actual trajectories can be compared.

Furthermore, the ZFEL does not claim that our random-walk model is the only possible one. We think it is a good one, and it is the one we have adopted here and in previous publications, but there may well be others. For example, the scale in Figure 1A is arithmetic, or implicitly logarithmic, as discussed, but other scales are possible (including some of higher dimensionality), and in certain contexts could be appropriate. Also, Figure 1A models change with a discrete-step random walk. An alternative is a Gaussian random walk, and we explore this later. But there are still other discrete-time functions that could be used, and presumably even various continuous-time functions. In sum, the central claim of the ZFEL is that some decomposition of diversity and complexity into ZFEL and non-ZFEL components is necessary. The manner in which the ZFEL component is modeled is a separate matter.

To be clear, most understood the ZFEL perfectly from the start. But these misunderstandings were common enough that we feel a special obligation to try to clarify further. The ZFEL is a frame of reference, one that studies of diversity and complexity have previously lacked. Imagine two people on a train playing catch with a ball, one tossing to the other in a forward direction, in the same direction the train is moving, and the other tossing it backward. To make sense of the ball's movements, relative to a frame of reference outside the train, it helps enormously to understand them as a combination of two independent motions, the motion imparted by the train and the motion imparted by the people making the tosses. Indeed, it is almost impossible to make sense of the ball's movement without making this decomposition. Returning to biology, substitute diversity or complexity for the ball, and the ZFEL for the train that imparts a directional tendency to them. Our claim is that it would be difficult to explain the actual trajectories of diversity and complexity without decomposing them into a ZFEL and non-ZFEL component.

To translate our earlier point into these terms, suggesting the movement of the ball can and should be decomposed into a train and nontrain component makes no empirical claim about the trajectory of the ball, about whether it will actually move forward or backward. To say that it is on a train that is moving in a particular direction is not to claim that its net movement will be in that direction. Nothing in this model says that the ball cannot have a net backward movement, relative to a frame of reference outside the train. In fact, if the backward tosses are faster than the train, it will be moving backward frequently. In biological terms, if natural selection opposed diversity or complexity strongly and frequently enough, diversity or complexity will often decrease. In principle it could be that opposing selection is so strong, pervasive and persistent that diversity or complexity *always* decrease over the history of life. It is pretty obvious that this is not the case. But the point is that the ZFEL is a statement about how to understand change in diversity and complexity, not a statement about how they actually change.

One final remark is needed here, to head off a worry that will occur to some readers when they get to Section 5. In that section, we make a number of empirical claims, and show how they follow from the ZFEL. So, for example, we claim that the vast majority of the divergence that has occurred in the history of life must have been driven by the ZFEL. This is an empirical conclusion, and the worry is that it might seem to contradict our insistence here that the ZFEL makes no empirical claim. In fact, it does not, because our argument introduces other facts of biology along the way, and it is these facts, in combination with the ZFEL, that makes claims like this possible. Analogously, Newton's laws by themselves make no empirical claim. By themselves they will not tell you how fast an object is moving. But Newton's laws plus some facts about initial velocity, mass, and applied forces will tell you exactly that. Likewise, the ZFEL makes no empirical claim, but the ZFEL in combination with other facts of biology can lead to empirical conclusions.

3.2 The ZFEL Is Not Empirical but Is Based on an Empirical Truth

If the ZFEL makes no empirical claim about the behavior of diversity and complexity, just what is its empirical content? Consider the law of large numbers. It tells us that if we perform a large number of trials (say, flipping a coin) the relative frequency of a result (say, heads) will be close to the mathematical expected value of the result. In this case, it tells us the #H/#Tosses $\cong$ P(H) when the number of tosses gets large. There is a mathematical theorem in the probability calculus that give us this result. And so, one might think that the question of whether or not it is true would be unnecessary – of course, mathematical theorems are true. But suppose you want to know what will happen tomorrow when you toss a coin with a known (highly verified) probability of heads 1,000 times. It is a simple matter to use the probability calculus to calculate error bars and so to say with a certain confidence level, say, 99 percent, that the result will fall within a certain interval. And that is that.

But now suppose at some point in your life you read David Hume on the problem of induction. Hume pointed out that inference to the unknown or unobserved is not justified by any deductive or inductive inference. All such inferences beg the question of whether or not tomorrow will be like all past days. It need not be, the world may go crazy tomorrow, and if it does, you better not trust the law of large numbers!

It may be puzzling that an analytic result of mathematics may be wrong. Consider this: all nourishing bread is nourishing. That has the logical form of "all A's that are B are B" and so cannot be false. But contrast that statement with this: bread that has nourished me in the past will nourish me tomorrow. That clearly is not analytic, and may well be false.

It turns out that we live in a world where what has been nourishing in the past is in fact a good guide to the future. More generally, we have discovered through empirical means that we live in a world where probability values measured carefully in the past are in fact a good guide to the future. To use Goodman's (1955) terminology, we live in a world where probabilities are *projectable*. But that, as Goodman pointed out, is an empirical fact about the world. Thus, the law of large numbers is an empirical truth about this world and the principle of direct inference (the principle that allows us to move from probabilities to frequencies [or data] is one that reflects a very deep truth about this world).

The ZFEL, both special and general versions, is based on this deep empirical truth about this world. In Section 4 we will present mathematical demonstrations of the ZFEL. We will show why and how independently evolving lineages diverge from each other with a quantified probability. That is math. But the application of that math to the world relies on the deep empirical truth about the projectability of probabilities.

So, there are two questions we can ask about the quantified versions of the ZFEL presented in Section 4. Is the math right? That is, are the equations we present mathematical consequences of the rather sparse assumptions we start with. Spoiler alert: the math is right. Second, we can ask what the application of this bit of mathematics to the biological world depends on. And the answer to that is that it depends on the deep, and welcomed, empirical fact that probabilities work in this world. The ZFEL is based on this empirical fact. And in that sense it is empirical, without making any empirical claim.

3.3 The ZFEL Is Not the Second Law

The Second Law of Thermodynamics was originally thought of as an empirical, even phenomenological, truth about the world. Carnot showed the impossibility of a perpetual motion machine, by showing the impossibility of getting as much usable energy out of a machine as had been put in. The heat lost in any such machine was labeled entropy by Clausius who gave us the first version of the second law, which basically stated that in any closed system entropy can never go down. How this law related to the rest of physics was addressed by the work of Maxwell and Boltzmann in the late nineteenth century. Their work produced the statistical-mechanical version of the second law, which is the one with which most readers are familiar. That version, unlike the first, is thoroughly probabilistic. It works by assigning probabilities to the micro-states (position and momentum of every molecule). Some micro-states are improbable, e.g., a gas in a cubic container where all of the fast-moving molecules are on the right and all of the slow-moving ones are on the left, while some are much more probable. If a system starts in an improbable state it will (probably) evolve into a more probable state. Unlike Clausius's law, here entropy in a closed system can decrease, it just hardly ever does.

The statistical-mechanical version of the second law can be derived in a very abstract and mathematical way (see, e.g., Penrose 2005). There is no physics involved. Similarly, one might think the derivations of the ZFEL in Section 4 are abstract, mathematical and involve no biology. So, what is the relation between the second law and the ZFEL?

We do not want to rule out the possibility that someone will come along and show some interesting relation between the ZFEL and the second law. But we want to warn against any quick and facile conclusions here, in particular, the conclusion that they are one and the same, or that the ZFEL is a special consequence of the second law. Our view is that they both have their origins in probability (see McShea and Brandon 2010, 109–11). However, here we want to briefly cover four important differences between them:

1   The ZFEL is while the second law is not, hierarchical. As discussed Section 2, there is no privileged level of analysis for diversity and complexity. Cells make up organs and tissues, which make up organisms. Organisms form groups and species. Species form genera, and so on. The ZFEL applies at all of these levels and more. Diversity and complexity are level relative. As a consequence of this it is noncontradictory to say that a diverse genus is composed of species of low diversity. Or that a complex organism has simple cells. (And in fact, as we discuss at the end of Section 5, cells in multicellular organisms are simpler than the free-living, single-celled protists they evolved from.) The analogous thing cannot happen with respect to the second law. A space cannot have low entropy while all of its subregions have high entropy, nor vice versa. If we measure the entropy of all of the parts, we have the entropy of the whole.

2   The ZFEL is, while the second law is not, intended to be a null model. We have already discussed how we intend for the ZFEL to be used as a null model. This means it does not make direct empirical claims about the world. The second law makes exactly that sort of claim.

3   The ZFEL is, while the second law is not, biological. Notice that the ZFEL is relevant only for evolutionary systems, that is, systems with heritable variation. But heredity implies reproduction, which puts us into the realm of biology. (It might well be applied to cultural systems too, but they are biological.) The second law applies to an internal combustion engine, and also to a container filled with oxygen. That is not biology.

4   The ZFEL and the second law paint very different images of the world. Given free rein, the ZFEL predicts a world filled with diverse and complex biological forms. As Darwin (1859, 425) pointed out, "There is grandeur in this view of life, with its several powers, having been originally breathed into a few forms or into one; and that, whilst this planet has gone cycling on according to the fixed law of gravity, from so simple a beginning endless forms most beautiful and most wonderful have been, and are being, evolved." If left to its own, the second law predicts a cold gray uniform wasteland.

Based on points 1–4, it seems safe to say that the ZFEL is not the Second Law.

4 The ZFEL Quantified

Two entities evolving independently will, probably, diverge. How much will they diverge and how likely is that divergence? Those are quantitative questions, and in this section we will answer them. (See McShea et al. 2019 for a more condensed presentation of the quantitative models developed in this section.)

Variation is itself variable. Sometimes it comes in discrete units, for instance arm number in sunstars, or number of black spots on a ladybug. And sometimes it is continuous, for instance height in humans or body size in deep-ocean ostracods. We need not insist on some fundamental ontological difference here – continuous variation can always be treated discretely by binning the resulting measurements. For instance, measured heights of humans can be put into a discrete number of bins. And discrete variation can be handled within a continuous model – a continuous model does not require all possible values be realized. We could treat height as continuous even if everyone in our population were either exactly 5 feet tall or 6 feet tall. But sometimes it seems most natural to treat variation discretely and in other cases the continuous treatment works best. Below we will develop a set of equations that answer our two questions – How likely? and How much? – in a system with discrete variation. Then we will develop another set of equations for the continuous case. We will then apply the latter set of equations to a real data set, to illustrate how this can be done.

One might argue that the continuous case is the more general and that therefore developing the mathematics for this should suffice. But we think that one virtue of developing the mathematics for the discrete case first is that this case is easier to understand. Some readers may find the mathematics of the continuous case too much. For such readers we advise paying close attention to the discrete case. Our hope is that we can make this case clearly understandable. Once understood, you may want to just trust us that the basic logic for the continuous case is really the same as that for the discrete case.

So one of our aims in this section is to clarify the how and the why of the ZFEL. The mathematics, we think, helps enormously in this. A second aim is just proof of concept. In our original book (McShea and Brandon 2010) we stated that the ZFEL should be quantified and saw no impediment to its quantification, but we did not do it. Now we have. A final, and perhaps most important aim is to provide a zero-force theoretical framework for evolutionary theory. Such a framework would allow hypothesis testing, in particular testing the ZFEL null, and hypothesis generation to go hand in hand.

4.1 ZFEL Quantification – the Discrete Case

We start with a very simple, and general, model of an evolutionary system. Two entities, **A** and **B**, sit on a one-dimensional scale. They can do one of three things: move left, stay put or move right. Each of these moves is governed by a probability, p_1, p_2, and p_3, where $p_1 + p_2 + p_3 = 1$. For simplicity we will assume that $p_1 = p_2 = p_3 = 1/3$, but that assumption can easily be relaxed. The moves occur in a single time step, which can be thought of as a generation (see Figure 2). In this setup, if the entities are populations or taxa, the difference in coordinate values between A and B is their disparity (i.e., diversity), and if they are parts in an organism, the difference is their degree of differentiation (i.e., complexity).

Now consider the joint moves of the two entities. There are nine possibilities. One of them we call *stasis*. *Stasis* is when no part of the system changes and so happens when both **A** and **B** stay put, and that happens with probability $1/3^2$. Whatever the initial distance between **A** and **B**, stasis leaves that distance unchanged. Two other possibilities also result in no change in the distance between A and B, namely, when the two move in parallel. That is, when both **A** and **B** move to the left, and when both move to the right. Thus of the nine possibilities of joint moves, three leave the distance between **A** and **B** unchanged. Another qualitative possibility is that this distance decreases. This happens when **A** moves to the right and **B** moves to the left, or when **A** moves to the right and **B** stays in position, or when **A** stays put and **B** moves to the left. We call this *convergent change*. The final qualitative possibility is that the distance between **A** and **B** increases, which we call *divergent change*. This happens when **A** moves to the left and **B** to the right, or **A** to the left and **B** stays put, or **A** stays put and **B** moves to the right.

4.1.1 An Incorrect Intuition

As we saw above, this system of joint moves of **A** and **B** has three qualitative possibilities, no net change (either *stasis* or *parallel* change), *convergent change,* or *divergent change.* Each of these three categories is realized by

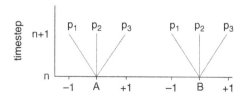

Figure 2 Two entities, **A** and **B**, moving in a one-dimensional space over one time step, both with probabilities of increase, stasis, and decrease equal to 1/3, 1/3, and 1/3, respectively.

three of the nine possibilities of joint moves. Given our simplifying assumption that the three transition probabilities are all identical (= 1/3) it follows that the probabilities of the three categories are also all identical. And so, *divergent change* is no more likely than *convergent change*.

Put more generally, consider two lineages evolving (independently) through morphospace. If change is random, then the probability of the two coming closer together is the same as the probability of them diverging. So, our expectation should be no net change in distance between them.

This notion is intuitively appealing and contradicts the claim of the ZFEL that two entities evolving independently will probably diverge. But it is wrong, and it is necessary for our purposes to show that it is wrong. But, more importantly, showing that it is wrong is essential for understanding just how the ZFEL works. The problem with this notion is that what happens over one time step, or one generation, cannot be generalized over many generations. Our two mathematical models presented below will demonstrate this conclusively. For now, let us offer a more intuitive explanation of this phenomenon, one that we will prove to be correct.

First, we want to mention one intuition that supports the idea that divergence is more likely than convergence. It is that there are more ways to get different than there are to get similar and so divergence is more likely than convergence. That is, one could build into one's model of independent evolution of multiple lineages the conclusion that divergence is more likely than convergence by simply specifying that in the model's initial transition probabilities. Such a model might even be useful in some cases, but not for our purposes. Our explanation of why divergence is more likely than convergence does not rely on this asymmetry in transition probabilities.

Instead, here is why there are more ways to diverge than converge. Consider two species of lady bugs that differ in the number of spots on their wing covers. Species **A** has eight fewer spots than **B**. Now suppose **A** moves toward more spots, say, one more spot, in the first time step and that **B** moves toward fewer spots, say, one less spot in one time step. Thus, the two species have become more similar with respect to spot number. This is convergent change. But now suppose those trajectories are maintained for a few generations. Eventually the two trajectories cross each other so that what once was convergent becomes divergent. In the Section 4.1.2 we will show how this happens precisely, but for now we will offer this as explanation: The qualitative change from convergent to divergent is asymmetrical: all consistently converging trajectories end up diverging, but no consistently diverging trajectories end up converging. And further, all trajectories that converge on average eventually diverge (after they cross), but the reverse is not true of trajectories that diverge on average. And so, across all possible pairs of trajectories, divergence predominates, even when movement is random.

4.1.2 The Discrete Model

Recall our four categories of joint change: *stasis* (S), *parallel change* (PC), *convergent change* (CC), and *divergent change* (DC). Those four categories are exhaustive, so

$$P(S) + P(PC) + P(CC) + P(DC) = 1.$$

Let NS stand for *nonstasis*. Nonstasis happens when any part of the system changes over any time period. So, rather obviously,

$$P(NS) = P(PC) + P(CC) + P(DC)).$$

Rearranging the first equation, we get

$$P(NS) = 1 - P(S).$$

A single entity, say, **A**, changes over a single time step if it moves to the right ($P = 1/3$) or to the left ($P = 1/3$), so the probability that it changes over one time step is 2/3. We will call this probability the *intrinsic rate of change,* and will label it r. In our simple model, **A** and **B** both have the same r. So for a single time step in this system of two entities, the probability of stasis is $(1 - r)^2$. Over t time steps, the probability of nonstasis is

$$P(NS) = 1 - (1 - r)^{2t}.$$

One important consequence of this is that as the number time steps, n, increases the probability of nonstasis quickly approaches 1. In our example, at $t = 1$, $P(NS) = 8/9$, or 0.889 to three decimal places; at $t = 3$, $P(NS) = 0.9986$; and by $t = 5$, $P(NS) = 0.99998$. Put another way, in this simple model, stasis is incredibly unlikely. (Note as well that the way we are defining stasis requires no change at any time, not merely no net change. No net change is not as unlikely as no change at all.)

We are not interested in change in general, but rather with divergent change. We can combine parallel change (PC) and convergent change (CC) into one quantity, which we will call *nondivergent change* (NDC). Thus

$$P(DC) = P(NS) - P(NDC).$$

Adding and subtracting P(S) on the right side of the above equation gives

$$P(DC) = P(NS) + P(S) - P(NDC) - P(S).$$

But change and stasis are mutually exclusive and exhaustive options in our system, and so P(*nonstasis, i.e., change*) + P(*stasis*) = 1. Therefore the above equation simplifies to

$$P(DC) = 1-[P(NDC) + P(S)]. \tag{1}$$

The term in the square brackets, the sum of the probability of nondivergent change plus the probability of stasis, can be computed directly by going through all of the possible combinations of moves of **A** and **B**.

To see how this is done, consider Figure 3. One feature of our approach is worth noting now. In Figure 3 the phenotype of interest – arm number in a sunflower starfish (also called sunstars) – is placed on a scale on which a distance can be measured. This can be thought of as a geometrical approach. We adopt it here and in the treatment of the case of continuous variation. We think this approach is quite general in that it applies to most phenotypic features, whether morphological, physiological or behavioral. To help fix ideas, we will focus on morphological examples, but the approach is not limited to such cases. It is limited, however, to cases where the feature of interest can be placed on a scale like this. Amount of pigmentation can be put on such a scale, but not color. (Wavelength or frequency lie on such a scale, but they are different from color.) Interestingly, we see no nonarbitrary way of putting the four DNA bases on such a metric scale. (What is the distance between A and T? Is it greater than, or less than, or equal to the distance between A and G?)[1] We would be pleased to be proven wrong on this point.

Figure 3 extends our model over five time steps or five generations. The figure shows two sunstar populations. Sunstars vary in arm number and here we assume that arms come in whole numbers and that they are added or subtracted by one per time step. This example is purely hypothetical and is meant to show how the model works, not to tell the reader facts about sunstars.

The one-way trifurcating lattices represent all of the possible trajectories **A** (lightly shaded) and **B** (heavily shaded). Let A and B denote the positions of lineages **A** and **B**, with A_0 and B_0 denoting their initial positions. Note that at generation t, lineage **A** can take on $2t + 1$ possible values: $A_0 - t$ through $A_0 + t$. Likewise for **B**.

For each possible move of lineage **A**, classify the moves of lineage **B** into two categories: divergent and nondivergent (the latter consisting of parallel, convergent, and stasis). The initial value of $B - A$ is $B_0 - A_0$; in our example we have $B_0 - A_0 = 1$. Since trajectories of **B** and **A** can cross so that $A > B$,

---

[1] Note that what is called genetic distance is defined in terms of the number of mutations required to get from one sequence to another, not in terms of a fixed metric.

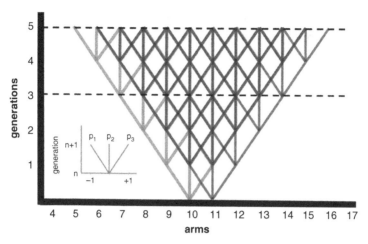

Figure 3 All possible trajectories for two entities, *A* (lightly shaded lines, arising at 10) and *B* (more heavily shaded lines, arising at 11), over five time steps. The inset reprises Figure 2, showing the probabilities of increase, stasis, and decrease at each time step, 1/3, 1/3, and 1/3, respectively. In the discussion, the entities are two populations of sunstars, and the horizontal axis is number of arms, presumed to be strongly heritable.

divergent moves are ones for which either $B - A > 1$ or $A - B > 1$, that is, for which $|B - A| > 1$.

To calculate the probability of divergence, we turn to Equation 1. Calculating the sum of the probability of nondivergent change plus the probability of stasis is straightforward. We start by taking a fixed value of A_i and then summing the probabilities of values of B that result in $|B - A| \leq B_0 - A_0$ (which in this example is 1). Then we move to the next possible value of A_i and repeat, and we continue until we have gone through all possible values of A. Expressed as a formula, the probability of nondivergence is as follows:

$$\sum_{i=A_0-t}^{A_0+t} P(\mathbf{A} = i) \sum_{|j-i| \leq B_0-A_0} P(\mathbf{B} = j).$$

This gives us a simple measure of all of the possible pairs of trajectories that leave $|B - A|$ at or below the original distance between them. That is, this is the probability of nondivergence. It includes all of the pairs that would fit into one of our qualitative change categories (convergent or parallel), but it also includes the single case of stasis. Recall that stasis means no part of the system changes at any time. Thus, stasis excludes a trajectory that, say, moves one unit to the left, then one unit to the right. (That case is one of no net change, but it is still change.) The trifurcating pattern of our model produces 59,049 $[= (3^5)^2]$ possible trajectories for **A** and **B** by generation 5, but there is exactly one of these (one out of 59,049) that is stasis, namely, when A stays at 10 and B at 11.

The double summation above gives us the probability of nondivergence, i.e., the probability that the distance between **A** and **B** remains at or below the initial distance between them (i.e., $|B - A| \leq 1$). Returning to Equation 1, we can calculate the probability of divergence as follows:

$$P(DC) = 1 - \sum_{i=A_0-t}^{A_0+t} P(\mathbf{A} = i) \sum_{|j-i| \leq B_0-A_0} P(\mathbf{B} = j). \qquad (2)$$

To illustrate how Equation 2 works, we will go through the cases of $t = 3$ and $t = 5$.

We have already seen how the model works over one generation: 1/3 of the time it results in convergence, 1/3 of the time it results in divergence and the final third is stasis. And we have pointed out that this set of results cannot be extended over generations. This is a good place to explain and reinforce that point. At time step 3, or generation 3, **A** has seven possible states, ranging from 7 to 13 arms. Similarly, **B** has seven possible states, ranging from 8 to 14 arms. To calculate the probability of divergent change we simply calculate the probability of nondivergence which we then subtract from 1. To calculate the probability of nondivergence we go through all possible combinations of these states seeing which result in nondivergence. We start at $A = 7$. At that point only one possible state of **B** results in $|B - A| \leq 1$, namely, $B = 8$. What is the probability that $B = 8$? To figure that out we need to trace every possible trajectory that lands at $B = 8$ at $t = 3$. It turns out there is only one such trajectory, namely, the one that moves one step to the left at each time step. Given that each such transition has a probability of 1/3 we now know the probability of $B = 8$ at $t = 3$. It is $1/3^3$.

Next we move to $A = 8$. At that point there are two possible states of **B** such that $|B - A| \leq 1$, namely, $B = 8$ and $B = 9$. Again, both of these states have a well-defined probability.[2] We continue until we reach **A** $= 13$. Adding all of these probabilities together gives us the probability of nondivergence. That quantity subtracted from 1 gives us the probability of divergent change. In the case of $t = 3$, it equals 0.5103.

---

[2] We have already seen that the probability of $B = 8$ is $1/3^3$. That is the probability of the single trajectory going from B_0 to $B = 8$. There are three trajectories that go from B_0 to $B = 9$. Here is where our assumption that the three transition probabilities, $p_1, p_2,$ and p_3, are all equal comes in handy. Because of that equality, every trajectory has the same probability. However, this is not at all necessary for the required calculations. Any given trajectory is a series of transitions (e.g., move right in step 1, stay put in step 2, move left in step 3), for which there will be a well-defined probability that is the product of the probabilities of each of the transitions.

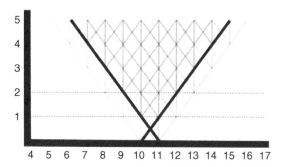

Figure 4 Reprise of Figure 3, highlighting the crossing of two extreme trajectories, the maximum for **A** (which started at 10) and the minimum for **B** (which started at 11).

Recall that at time step 1 the probability of divergent change is 1/3. At time step 2 it is 4/9. It is not until time step 3, or generation 3, that this quantity exceeds 0.5. By generation 5, $P(DC) = 0.5924$, and it continues to increase with each generation. How does what is unlikely at the first time step become more and more likely? It does so because trajectories of **A** and **B** cross so that what started as convergence becomes divergence. To see how this works, consider the inner envelopes of the **A** and **B** distributions (Figure 4). In the first time step, these lines cross, leaving the distance between A and B unchanged. But by time step 2, $A_2 = 12$ while $B_2 = 9$, so that the distance between them is now 3. And thus this change is divergent. The reverse cannot happen, i.e., trajectories that start out divergent and persist divergently do not crossover to become convergent. This asymmetry explains why the intuitively appealing intuition discussed above is incorrect, and why divergence always wins out over convergence (probabilistically speaking).

There is a second aspect of these spreading lattices of possible positions that is quantitative. Again consider the shaded lines of Figure 4. By generation 2, they are three units apart, and so divergence has occurred, and following those line further does not change that qualitative result. But quantitatively the distance keeps increasing, e.g., by generation 5, they are nine units apart. So divergence wins out over convergence because of the asymmetry between them, and with passing generations divergence increases in magnitude.

Thus far we have quantified the likelihood of divergent change, i.e., ZFEL change. But we can also quantify the magnitude of such change. That is, we can also give a quantitative answer to "How much?" We do this by calculating the expected value of $|B - A|$. Building on what we have already done, this is fairly straightforward. We simply go through every possible combination of values of A and B, finding $|B - A|$ and then weighing that value by the appropriate probability. This is shown in Equation 3:

$$E|B - A| = \sum_{i=A_0-n}^{A_0+n} \sum_{j=B_0-n}^{B_0+n} |j - i| P(A = i) P(B = j). \tag{3}$$

At time step 1, $E|B - A| = 1.222$. Recall that the probability of divergent change in time step 1 was only 0.333, so one might wonder how this expectation could be greater than the original difference of 1. This happens because the distribution of values of $|B - A|$ is skewed. Already in time step 1 the distance between B and A can be as high as 3, but $|B - A|$ can never go below 0, thus causing a right skew. By time step 3, $E|B - A| = 1.774$, and by time step 5, $E|B - A| = 2.199$. The expected value of $|B - A|$ continues to rise with time. This gives us a quantitative null expectation against which real world data can be compared.

One might be interested in how Equations 2 and 3 behave over larger time-spans. We can give some results:

Generation 10	$P(DC) = 0.6921$	$E|B - A| = 3.016$
Generation 100	$P(DC) = 0.8960$	$E|B - A| = 9.238$
Generation 1,000	$P(DC) = 0.9675$	$E|B - A| = 29.210$
Generation 10,000	$P(DC) = 0.9892$	$E|B - A| = 92.187$

Both Equations 2 and 3 work by going through all possible combinations of A and B values. The combinatoric possibilities explode with generational time. Even at generation 10 there are 3^{10} possible trajectories of **A**, and similarly for **B**, thus $(3^{10})^2$ possible combinations, exceeding the memory capacity of most modern computers. (The above numbers come from an approximation procedure developed by our colleague Steve Wang.)

Before moving on to the continuous case, let us address one possible objection to the results given. One might complain that we have exaggerated the effect of the ZFEL by starting our two populations close to each other. It is true that the closer the two are in starting position the greater the likelihood of divergence and the greater the expected value of divergence, and we have already explained why this is so. The ZFEL effect requires the crossing of lines (see Figure 4), and the closer the two populations the sooner lines will cross. However, so long as there is a finite distance between the two initial positions, crossing will eventually occur. That is pure theory. But we want to make a second point that has to do with the application of this theory. We think that the most interesting and general applications of the ZFEL will be to recently separated parts (complexity) or lineages (diversity). Our theory shows that the initial driver of differentiation between previously identical things can well be the ZFEL. To explain such differentiation in other terms requires that one first rejects the ZFEL null hypothesis. Otherwise you are in the position of ignoring a completely general process that is capable of explaining your data, which is a bit like touting the miraculous powers of your new drug which, you claim, cures the common cold in about 10 days.

4.2 ZFEL Quantification – the Continuous Case

Arm number in sunstars might change discretely, but arm length changes continuously. The same goes for a huge range of organismal traits, from the dimensions of parts to the rates of biological processes to the concentrations of biological substances. Here we show how to calculate, paralleling the discrete case, the probability as a function of time that two continuously varying entities will diverge and how to calculate the expected value of that divergence, that is, the magnitude of the expected increase.

Consider two entities, **A** and **B**, starting at different values along some continuous axis of variation, as shown in Figure 5. We have arbitrarily set the starting points for **A** and **B** at 2.0 and 3.0, respectively. One can think of the entities as two species, in which case their initial difference – or their disparity – is 1.0. Or one can think of them as two parts in a single individual, with the axis representing some body part dimension like length, in which case their initial complexity is 1.0. For purposes of illustration, let us assume that these are species, and this is a body-size axis. Further suppose that changes in body size are proportional and therefore additive on a log scale. What that means is that, for example, if a change in size of one unit on the scale represents a doubling of size, then a change of two units represents a fourfold increase, three units represents an eightfold increase, and so on. Also, all of the values on the log scale in Figure 5 are positive, but the scale is understood to extend without limit in both directions, so that negative values simply represent very, very small body sizes.

Here, increases and decreases over time are not drawn from a discrete distribution like the trifurcating (1/3, 1/3, 1/3) distribution in Figure 2, but from continuous distributions – called step size distributions– that we position symmetrically over the entities' starting points, as shown in Figure 5. To make computation easier, we use normal distributions in the figure, labeled "a" and "b" (within circles). To calculate the size of each entity at time n+1, we draw a value from its step size distribution and add it to the entity's value at time n, that is, to 2.0 and 3.0 for **A** and **B**, respectively. The step size distributions have a mean of zero, so the size of each entity will increase half of the time and decrease half the time. For a normal distribution, the mean is also the most likely outcome, so the most likely first step for both will be no change at all. That is, if

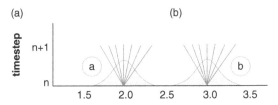

Figure 5 Two entities, **A** and **B**, moving in a continuous one-dimensional space over one time step. Here the moves of the entities are distributed continuously, following a normal distribution of outcomes (for **A**, the distribution is labeled "a" within a circle; for **B**, "b" within a circle). In both, the most probable step is no change, with **A** ending up where it started, at 2.0, and **B** ending up where it started, at 3.0. The angled line segments show some of the other steps the entities could take. Since the distributions are normal and continuous, all step sizes are possible in principle, although very large ones are improbable.

A starts at 2.0, the outcome that is more likely than any other is stasis at 2.0. Furthermore, most step sizes will be small. The fan of angled line segments above A in the figure shows the range of possible steps that fall within about one standard deviation of the starting point at 2.0, which for a normal distribution is the range of outcomes that can be expected about two-thirds of the time. However, we should point out that because the step size distributions are continuous, step sizes of any magnitude are possible in principle. The two tails of the step size distribution centered at A (lowercase "a" within a circle) may look like they taper to zero just short of 1.5 on the left and 2.5 on the right, but in fact those tails extend infinitely in both directions. A consequence is that **A** and **B** could in principle cross at any time, so that at time step $n + 1$, **A** could be larger than **B**, although with the step size distributions shown this is unlikely.

4.2.1 Probability of Divergence and Expected Divergence

Suppose we take the initial setup in Figure 5 to be the starting point of an evolutionary sequence, shown in Figure 6. The initial disparity between **A** and **B** is the difference between the two at time 0, or $3.0 - 2.0 = 1.0$ (in log units). Let us call this initial distance D, also as shown in Figure 6. And now we ask what is probability of divergence, the probability that the difference will be greater than D, greater than 1.0, after n steps? Likewise we can ask about the expected magnitude of the difference after n steps. To compute these, we extend the model in time, as follows.

First, to compute an entity's size at time 1, we draw a value from the step size distribution and add it to each entity's value at time 0. To compute it at time 2, we draw a value from the same step size distribution and add it to each entity's value at time 1. Continuing this procedure, each entity follows a random walk.

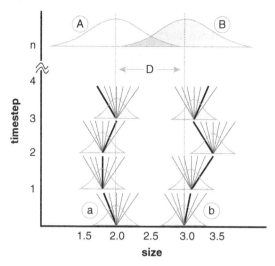

Figure 6 Trajectories of two lineages, initially distance D apart, both following Gaussian random walks. At each time step, each lineage changes by a value drawn from a normal step size distribution (circled lowercase a and b). The resulting distributions for each after n time steps (circled uppercase A and B) are shown above.

More precisely, it is called a Gaussian random walk, because the step sizes are not discrete but are drawn from a normal distribution. The heavy line segments in Figure 6 shows one possible trajectory for the entities over four time steps. Extending the run to n time steps, the cumulative results are the two large normal distributions (uppercase A and B within circles). The means of these large distribution are the same as the initial values, 2.0 and 3.0.

It is worth remarking here that from the unchanging means of the two lineages, one might be tempted to conclude that the expected difference between them will not change, that it will remain at D (i.e., 1.0). But as for the discrete case, this is wrong. And again, the reason has to do with the asymmetry discussed for the discrete case. When **A**'s net movement is to left and **B**'s net movement is to the right, the result is always divergence. But when the reverse occurs, when **A**'s net movement is to the right and **B**'s net movement is to the left, the result is not always convergence. Instead some of these will produce crossings that leave **A** more than 1.0 units to the right of **B**, adding to the net divergence total. Notice that here in the continuous case, the asymmetry arises immediately, in the very first time step. No matter how great the initial separation between lineages, in some tiny fraction of cases, the first steps will be large enough and in the right direction to produce crossing. In other words, in the continuous case, divergence is the expectation in the very first time step, with no lag, and forever thereafter.

We now proceed to the computation of the probability and magnitude of divergence, here broken down into four steps.

Step 1: Extending to n Time Steps. Consider just **A**. In the first time step, the range of possible locations for **A** expands by an amount that depends on the variance of step size distribution, σ_{ss}^2. With each subsequent time step, the range of possible locations for **A** increases. That is, the variance in **A**'s location increases, and it turns out that this increase is linear, so that after n time steps the variance for the location of **A** is $n\sigma_{ss}^2$. This is the variance of the large terminal distribution labeled A (with a circle). Similarly, if **B** has the same step size distribution, with variance σ_{ss}^2, then the variance in locations for **B** after n time steps is also $n\sigma_{ss}^2$. In Figure 7, the distribution for **B** is labeled B (with a circle). Writing this as an equation,

$$\sigma_n^2 = n\,\sigma_{ss}^2, \tag{4}$$

where, σ_n^2 is the variance of the terminal distributions, A or B, for the random walks after n steps, σ_{ss}^2 is the variance of the step size distribution, and n is the number of time steps. Two assumptions are built into Equation 4. One is that the step size distributions for both entities are the same. The other is that those distributions do not change with time. Both could be relaxed, at the price of making the problem more complicated.

Step 2: Divergence. We can now calculate the distribution after n steps of differences between **A** and **B**, in other words, the distribution for $B - A$. It turns out that because the distributions of **A** and **B** are normal and independent of each other, the distribution of $B - A$ is also normal, with mean equal to the mean of B minus the mean of A (which is the same as the initial difference, D or here 1.0), and variance equal to the sum of the variances of the A and B distributions, $2\sigma_n^2$. Figure 7A shows this distribution for $B - A$, centered at 1.0.

Of course, it is not $B - A$ that we want, but $|B - A|$ This is because, as discussed in the previous section, we need to include divergences arising when the two lineages cross, cases in which B ends up less than A by an amount greater than the initial difference, D. Figure 7B shows the first step in calculating the distribution for $|B - A|$. The $B - A$ distribution in Figure 7A is folded over at the zero line. In effect, the negative differences for $B - A$ are converted to positive differences, in other words, to absolute differences.

Step 3: The Probability of Divergence. We can now calculate the probability of divergence. It is the total area of the curve in Figure 7B that is to the right of D (1.0, vertical dashed line). Notice that this area is the sum of two pieces, the large right-hand half of the original $B - A$ distribution, which has an area of 0.5, plus the portion of the folded left tail of the original $B - A$ distribution that is to the right of D.

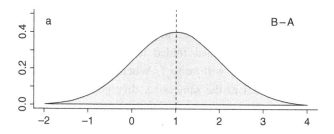

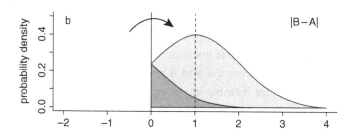

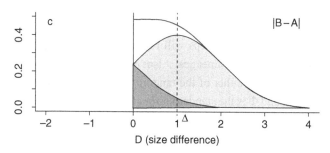

Figure 7 Distribution of the difference and absolute difference between two lineages, **A** and **B**, after *n* time steps. (A) Distribution of difference between **A** and **B**, where **A** and **B** are initially separated by one unit. Dotted vertical line indicates initial separation. (B) The distribution of absolute differences is generated by folding the distribution in A about the vertical line at $x = 0$. (C) The probability density function for $|B - A|$ (uppermost heavy line) was obtained by summing the curve bounding the lightly shaded region and the curve bounding the heavily shaded folded region. The triangle denotes the expected value of $|B - A|$. Note that this expected difference is greater than the initial difference of 1 (the dotted vertical line).

Two features of this result are worth pointing out. First, consistent with intuition, the smaller the initial absolute difference between the entities, the larger the size of the residual folded left tail, and the closer the total probability of divergence will be to 1. When the initial difference is zero, when the entities start at the same size, divergence is essentially certain. Starting from zero disparity or zero complexity, there is nowhere to go but up. Second, Figure 7B shows graphically and perhaps makes more intuitive the crossing effect discussed above. The portion of the folded left tail that lies to the right of *D* is the portion of the divergence that is due to the crossing effect. If we did not include this effect – in other words, if we excluded the folded left tail of the distribution – the probability of divergence would be exactly 0.5 and equal to the probability of convergence. But including the effect – that is, including the area of the folded left tail – the probability of divergence is guaranteed to be greater than 0.5. And this is true no matter how far apart **A** and **B** start, and no matter how small the variance of the step size distribution. Even at the very first time step, some nonzero sliver of the folded left tail will extend into divergence territory, increasing the probability of divergence above 0.5.

We can now give an equation for the probability of divergence:

$$P(DC) = P(B-A) > D + P(A-B) > D = 0.5 + \left[1 - \Phi\left(\frac{\sqrt{2}D}{\sigma_n}\right)\right], \quad (5)$$

where P(DC) is the probability of divergence, or the probability that the distance between the lineages after *n* steps will be greater than *D*, the absolute initial difference between the two lineages. Also, σ_n is the standard deviation of the terminal distribution of either of the random walks (which from Equation 4 is equal to $\sqrt{n}\,\sigma_{ss}$), and Φ is the standard normal cumulative distribution function. Notice that this equation is the analogue of Equation 2, which gives the probability of divergence in the discrete case.

Step 4: The Magnitude of the Divergence. Finally, we can calculate the expected value of the |*B* − *A*| distribution, in other words, the expected absolute difference between A and B after *n* steps. The |*B* − *A*| distribution belongs to a class of well-known folded-normal distributions. Figure 7C shows the distribution after folding. The expected value can be calculated as

$$E|B-A| = \frac{2}{\sqrt{\pi}}\sigma_n * \exp\left(\frac{-D^2}{4\sigma_n^2}\right) + D\left[1 - 2\Phi\left(-\frac{D}{\sqrt{2}\sigma_n}\right)\right], \quad (6)$$

where $E|B - A|$ is the expected absolute difference between A and B and σ_n is the standard deviation of the terminal distribution of either of the random walks (which again is $\sqrt{n}\,\sigma_{ss}$). Also, D is the absolute initial difference between the two lineages, and Φ is the standard normal cumulative distribution function. Consistent with the prediction of the ZFEL, $E|B - A|$ increases monotonically with time, or in other words, the expected value of the absolute difference is always greater than the starting distance. In other words, the expectation is always divergence, an increase in disparity or complexity. Notice that this equation is the analogue of Equation 3, which gives expected values in the discrete case.

4.2.2 A Demonstration

Here we give a demonstration, using paleontological data, of the application of the equations. The data come from the work of Hunt et al. (2010), who used them in a study of body-size trends in deep-ocean ostracods, tiny crustaceans that secrete clam-like shells. Here the data are repurposed for a demonstration of the application of the equations above to disparity.

Possible Analyses. In principle, the equations above can be applied to any pair of lineages for which there is an overlapping time series of morphological measurements. That is, the two lineages must overlap over some temporal range, and morphological data must be available for both at some moment in time that can be designated a starting point, so that an initial separation, D, can be computed. Then, the equations can be used predictively. What is the probability that two zoo populations living under much-reduced selection will diverge and how much divergence is predicted? Equation 5 can be used to calculate the probability that the separation after the passage of some chosen amount of time will be greater than D, with time expressed in units of generations, n. And Equation 6 can be used to calculate the expected magnitude of the separation at that same time. This sort of predictive analysis could be done with the Hunt et al. data, but we do not do it here.

Alternatively, when the right data are available, the equations can be used as a selection detector. In particular what is needed are lineage pairs for which one has data at both an earlier time and a later time, so that one can compute an initial separation, D, and a later separation, a kind of before-and-after comparison. This is what we demonstrate below. The ZFEL always predicts divergence, but when the divergence is significantly greater than expected due to the ZFEL alone, that suggests that selection has been acting to drive the lineages apart. This can occur, for example, when two species compete for the same resources, and selection favors variations in either that make it different from the other, in other words, variations that reduce competition. The process is called competitive displacement, and we discuss it further in Section 5. Here we show how to detect this sort of selection using the equations. For any lineage pair, the actual separation at the common end-time can be compared to the expected, computed using Equation 6, and the significance of the difference between actual and expected can be calculated.

A special feature of the Hunt et al. data make another sort of demonstration possible. Hunt et al.'s data are body-size measurements. They found an increasing trend in ostracods over the past 40 million years, and concluded that it was driven by deep-ocean cooling over much of that period. However, they also identified a 16 million year period during which the cooling paused, the trend disappeared, and the pattern of change in ostracod lineages was consistent with a random-walk model. Here we do not test this conclusion, we assume that it is true, that the lineages are in fact following random walks, and that their trajectories are governed only by the ZFEL. What making that assumption does is give us an opportunity to show that the frequency of divergence does in fact follow the probabilities predicted by Equation 5 and the divergences predicted by Equation 6. We get to see Equations 5 and 6 in action, so to speak.

Computation of Critical Variables. Equations 5 and 6 have only two variables, D and σ_n. The initial separation D is assumed to be measurable directly. And σ_n is the standard deviation of the terminal distributions A and B (within circles) at time n. This can be computed using Equation 4 from the variance of the step size distribution ($\sigma_n = \sqrt{n}\sigma_{ss}$). There are various ways to estimate σ_{ss}. The first is to measure it directly in modern populations of the two lineages. Variable σ_{ss} is the standard deviation of the chosen morphometric variable in a set of offspring derived from a single pair of parents, or an average over the sets of offspring of many parents. Of course, for extinct lineages, generation by generation data are rarely available, and in those cases the step size distribution for a modern species in the same group might work as a proxy. A necessary assumption in both cases is that in a single generation, selection for divergence among individuals is either absent or negligible. (In the demonstration below, we took advantage of special circumstances and used a third method to estimate σ_{ss}. This, along with the rest of the details of the calculations, is explained in the supplemental information for McShea et al. [2019]. Here we just discuss the results.)

Selection Detection in the Ostracod Data. Figure 8 shows body size changing over time in 11 ostracod lineages. Many of the pairs of lineages have data points with the appropriate temporal overlap to produce a before-and-after comparison. For example, we could look at lineages **A** and **B** over the period from

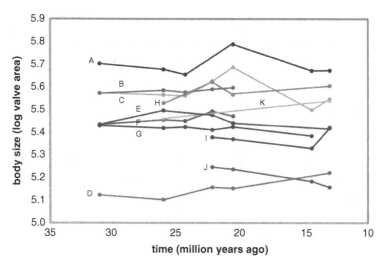

Figure 8 Body-size trajectories for deep-ocean ostracods. Data are from Hunt et al. (2010). See text for discussion. See the supplemental information for McShea et al. (2019) for species identification and for the original data.

26.12 million years ago to 24.39 mya. As the figure shows, body-size data are available for both lineages at both times.

The left side of Table 1 shows results over a 5 million year span for four lineage pairs (listed in column 1). These pairs were chosen because their starting points are close together (column 2). The ZFEL predicts that lineages starting close together will diverge quickly. We used Equation 5 to calculate the probability of divergence due to the ZFEL alone (column 3). All of the probabilities are quite high, over 90 percent, as expected from the ZFEL. And in fact after 5 million years, they all diverged (column 4). However, a close start means they began at a similar size, which raises the possibility that they could be competing with each other and that their subsequent divergence could also be partly the result of selection-for-difference. Column 5 shows the disparity predicted by Equation 6 after 5 million years for the ZFEL acting alone. Column 6 shows the actual disparity after 5 million years, and column 7 show the *p*-value for the difference between actual and expected. Two of the pairs diverged more than expected (EF and EG) and two diverged less than expected (BC and FG), and none of the differences from expectation were statistically significant. That is, all of the p values in column 7 were greater than the standard significance cutoff, 0.05. Thus, the data are consistent with (although not conclusively demonstrative of) the hypothesis that the pattern is the result of the ZFEL alone.

Table 1 Actual and expected absolute differences within selected lineage pairs

Lineage pair	Absolute difference	p(div)	div/conv	Expected absolute difference	Actual absolute difference	p**	Lineage pair	starting absolute difference	actual abs diff (d/c) (expected)	actual abs diff (d/c) (expected)	actual abs diff (d/c) (expected)	actual abs diff (d/c) (expected)
(1)	(2)	(3)	(4)	(5)	(6)	(7)	(8)	(9)	(10)	(11)	(12)	(13)
	31.12 mya=start	@26.12 mya	@26.12 mya	@26.12 mya	@26.12 mya		Lineage pair 26.12 mya	26.12 mya	24.39 mya	22.27 mya	20.63 mya	13.06 mya
Initial close spacing							initial moderate spacing	starting absolute difference				
EF	0.001	0.982	div	0.03600	0.042	0.35	BC	0.020	0.016 (c) (0.027)	0.036 (d) (0.036)	0.090 (d) (0.041)	
BC	0.002	0.965	div	0.03603	0.020	0.66	EH	0.033		0.146 (d) (0.042)	0.126 (d) (0.047)	0.185 (d) (0.064)
EG	0.004	0.930	div	0.03613	0.078	0.08	CH	0.035		0.004 (c) (0.043)	0.120 (d) (0.048)	0.059 (d) (0.065)
FG	0.005	0.912	div	0.03621	0.036	0.43	FG	0.036		0.082 (d) (0.044)	0.046 (d) (0.048)	
							EF	0.042		0.017 (d) (0.048)	0.030 (c) (0.052)	

Another feature of the left side of Table 1 is worth a remark. As the initial absolute difference increases (EF to BC to EG to FG), the probability of increase – that is, the probability of ZFEL-driven divergence – decreases (column 3). The reason is that for any fixed time interval, as the starting distance between lineages increases, the probability that they will cross decreases, and as discussed earlier, it is the crossing that produces the ZFEL divergence. In the extreme case, where the initial separation is very large, the probability of crossing would drop to near zero, and the ZFEL expectation would be that the terminal difference will be nearly the same as the initial difference, D. Still, as pointed out earlier, the probability of divergence will always be greater than 0.5, and the expected divergence always greater than D, however minimally.

The Equations in Action. The right half of Table 1 shows actual and expected changes in diversity as a function of time for six lineage pairs. Here we assume that the data represent a system of pure random walks, as Hunt et al. suggest, and the point is to show how the expected probability of divergence (from Equation 5) and the expected magnitude of divergence (from Equation 6) increases over time. The starting point for all pairs was 26.12 mya, at which time the members of each pair were a moderate distance apart. For all pairs, the data set is complete enough to show changes in their disparity at a number of times over a relatively long span. As expected in a pure-ZFEL system, the frequency of divergence increased with time: at 22.27 mya, there were three divergences and two convergences; at 20.63 mya, there were four divergences and one convergence; and at 13.06 mya, both pairs for which data were available were divergences. Also, consistent with a central claim of the ZFEL in the continuous case, divergence is the expectation for all pairs at all time-spans. That is, all expected values (parentheses) are greater than the starting distances.

4.3 Methodological Matters

We have argued that the ZFEL is the appropriate null model for all investigations into diversity and complexity. Why is it appropriate? Because, as we have shown, without anything else intervening, the ZFEL process produces differentiation, i.e., diversity and/or complexity. Selection, or other evolutionary forces, might augment or oppose the ZFEL, but in order to demonstrate that one needs to know what the ZFEL baseline is. In this section we have shown how to produce that baseline both in discrete and continuous cases.

Therefore, the first step in any evolutionary study of diversity/complexity needs to be a testing of the ZFEL null. But after that step, we say let a thousand flowers bloom. That is, we are not advocating an exclusive use of ZFEL null model testing. Suppose you fail to reject the ZFEL null, as in the ostracod data discussed above. Logically that does not entail that the ZFEL is true. So, one then might want to do further studies, perhaps involving maximum likelihood testing, to see whether one ought to accept the ZFEL in that case.

In general, rejecting a null hypothesis doesn't get one much. But, as discussed in Section 2.1, in the context of a zero-force theory, rejecting a null leads pretty directly to alternative hypotheses. In Newtonian physics if we see a planet-sized object accelerating, we can, by Newton's first law, know that a force is acting on it, and know how to quantitatively characterize that net force. Newton's first law does not, however, tell us the source of the force. But there are only so many Newtonian forces out there and none but gravity are going to be effective at the planetary scale. So, we are led to a gravitational hypothesis. Similarly with respect to the ZFEL. If two lineages differentiate much more than would be expected by the ZFEL alone, nothing much but selection could possible explain that. In contrast, if two lineages differentiate much less than the ZFEL expectation, we could explain that by either selection or by some sort of constraint. Further investigations would be needed to decide which option is more likely.

5 What the ZFEL Means for Biology

Two entities evolving independently will, probably, diverge. That is the ZFEL. As shown in the previous section, how likely it is that they will diverge and the amount they will diverge are quantifiable matters. For now, we want to draw out the ramifications of the fact that it takes two to tango.

Imagine a world with a single species. In that world the ZFEL would not operate on the species level. (However, it would operate at various subspecific levels as our theory is hierarchical, that is, the theory would apply at whatever level two or more independently evolving entities are found.) Conversely, in a world with multiple species the ZFEL-based dispersion of species would not be understood if we were to look solely at single species' trajectories through morphospace. In this sense, the ZFEL is a second-order phenomenon, or an emergent phenomenon, one that emerges from the relationship between two or more entities

In this section we will show that the sort of emergence involved here is not at all mysterious, even though some philosophers are needlessly afraid of the notion. Next, we will examine extant explanations of diversity and argue that although there is much left to be empirically investigated here, at present the best explanation for the vast majority of cases of disparity is in fact the ZFEL explanation.

The case for complexity is interestingly different. We have argued that diversity and complexity are the same thing viewed at different hierarchical levels. So, one might think that since we hold that the ZFEL is in fact the best explanation for most cases of diversification we are committed to the same position with respect to complexity. We are not and will explain why not below. We will argue that we have good empirical evidence showing that the ZFEL is not the complete story with respect to complexity, that quite often selection acts against complexity. That generalization seems sound, but there is enormous room yet for empirical investigations of particular cases.

Sections 2–4 of this Element have shown how the ZFEL produces increases in diversity and complexity in the absence of constraints or forces opposing diversity and complexity. Because of that fundamental property of evolutionary systems, we have argued that one must start with the ZFEL null model when studying the evolution of diversity and complexity. All of this is independent of the particular ways in which evolution has unfolded on this planet (or elsewhere). That is what we meant when we have emphasized the the ZFEL itself makes no empirical claims (see in particular Section 3.1). But in what follows we will be using the ZFEL, in combination with some facts about biology, to make some empirical claims about just how diversity and complexity have evolved on this planet.

5.1 Emergence

Emergent phenomena and emergent objects are all around us and only people who have read too much philosophy fail to see this. Life emerges in certain arrangements of nonliving matter. Corporations emerge in complex human economies. Sometimes the properties of higher-level entities are of the same kind as the properties of their parts, and sometimes emergent properties are different in kind. For instance, the mean height of a population of plants is a property of the population, not of the individuals that compose the population. However, mean height is measured on the same scale as individual height and is not, therefore, different in kind. In contrast, the variance in height needs to be represented on a different scale, a second-order scale. Complexity and diversity are, as we have seen, variance concepts.

Mean and variance are mathematical concepts, but the same point can be made with respect to nonmathematical concepts. Human rights emerge in certain types of human legal systems, but no cell or organ has rights. In contrast, a flock of birds has an aerodynamic shape, different from the shape of any individual bird, but not different in kind.

Here we are not interested in a precise explication of the concept of emergence, the commonsense understanding of the term suffices for our purposes.

Rather we are interested in the epistemological (or methodological) and ontological consequences of emergence.

We start with a reasonably well-understood case of emergence, the case of drafting at NASCAR superspeedways. Aerodynamic drag increases with the square of velocity (approximately). Thus, drag becomes more of an issue at the speedways where the highest speeds are achieved (Daytona and Talladega). Drafting is when one car gets within inches of the rear bumper of another car. The result is that both can go faster. At Talladega drafting cars go 3–5 miles per hour faster than they can traveling alone. The drag acting on a single car results from the front being pushed back by the air through which it is traveling and from the rear being pulled back by the low-pressure area formed at the rear of the car. A pair of cars drafting splits these two components of drag, with the front car experiencing the push from the air, but not the low-pressure pull at its rear. Only the trailing car experiences this pull, but it escapes the push on its front – the air has already been split for it.

We want to draw two lessons from this example. First, from a methodological point of view, we would fail to correctly predict the average lap speed of a car if we simply looked at what the car can do in isolation. This is true whether we are studying this purely empirically, i.e., just measuring lap times with a stop watch, or studying it theoretically where we write an equation based on horsepower, weight, shape and size that predicts speed. For example, a particular car at a particular track may have a maximum average lap speed of 195 mph. But in a race where it is drafting, it has laps where it achieves an average lap speed of 199 mph. The speed of 199 mph is emergent and not predictable based on the behavior and capacities of the car in isolation.

Second, this case involves an emergent object, which is the pair of cars. We are not permitted to choose whether we recognize pairs of cars as objects based on our philosophical tastes. Nature picks out the pair as a new aerodynamic object whose shape and size determine drag. And so, at least with respect to aerodynamics, the pair is real (and the individual cars are mere parts of the whole).

(For our readers who do not like NASCAR, one can make the same point with respect to bicycle racing. But the aerodynamics of a 60-person peloton is a bit more complicated than that of two cars.)

The ZFEL operates on collections of evolving entities. If we only look at each lineage in isolation, we will miss the ZFEL process and therefore miss much of what produces the diversity and complexity in nature.

5.2 Diversity

As discussed in Section 2, the most common measures of diversity in biology are simple counts of the number of different things within the larger collective, e.g., the number of species within an order. This has the virtue of being easily operationalized, but fails to address the question of how different are these different entities. As mentioned in Section 2, because of this limitation, a more informative concept was introduced into paleobiology, namely, that of disparity. Disparity puts the entities on some scale which allows one to address the question of how different. As we saw in Section 3, our quantifications of the ZFEL explicitly and quantitatively address that question. Thus, our primary interest is in disparity, but we hasten to add that this is really what biologists have been interested in since the time of Darwin. That is, when we ask whether or not diversity has increased over the history of life, we might initially address that in terms of counts of species, but only because we think that is a decent proxy for the spread of life through some morphospace. In this section we will be careful to distinguish the technical meanings of *diversity* and *disparity*, but when talking about the process, we will use the term *diversification*, which should not be identified with the narrow concept of diversity as a species count.

The ZFEL predicts and explains the process of diversification. In this section we will argue that for the overwhelming majority of cases of diversification the ZFEL is the best explanation. But that argument requires us to lay out the alternatives. We will cover three, the first of which was Darwin's. The next two are distinct logical possibilities which we feel the need to cover, even though, to the best of our knowledge, no one in biology or in the philosophy of biology has advocated for them.

5.2.1 Competition Leading to Character Displacement

In Chapter 4 of *Origin of Species,* Darwin puts forward what he calls the principle of divergence. Fleming (2013) argues that there are two distinct processes lumped together under this principle. The first is this: environments change, new resources (new niches) open up, and populations evolve independently to better fit their local niches. This just is the ZFEL explanation (or more precisely, it is a part of the ZFEL explanation – more on this in Section 4.2.3). Recall that the ZFEL only requires that lineages evolve independently of each other, in other words that their trajectories move randomly with respect to each other. The concept of randomness-with-respect-to is a second-order concept and cannot be reduced to the (absolute) randomness of its parts. Two lineages moving with absolute randomness through morphospace will in fact also be moving randomly with respect to each other. But the stronger notion of absolute randomness is not required for the ZFEL.

The second process is competition where an environment with fixed resources evolution "pushes" lineages apart so that they compete less and so more efficiently exploit these resources. This is competition in the modern ecological sense (Pfennig and Pfennig 2012). Presumably, this is the story behind the diversification of Darwin's finches on the Galapagos Islands (Grant and Grant 2006, 2008). Whatever Darwin thought about diversification, the idea that ecological competition explains all of, or at least a large portion of, the disparity we see in the biological world is one of the dominant views today. We will label this the "Darwinian" view (again without any commitment to what Darwin actually thought). That view is the subject of this section.

The competition hypothesis is an ecological hypothesis. Within a common environment two populations compete for a common limited resource. By definition, a competitive interaction between two populations has a negative effect on both (Pfennig and Pfennig, 2012, 3). This does not require symmetry in effects, for example one population may be large and well established while the other is small and new to the area. In this case the negative effect of the first population on the second will be larger than that of the second on the first. But both will be negative.

Here is the prototypical way this process is supposed to work. Some event splits an old population into two, say, a population of finches on an island splits. Because they were originally the same they are now very similar, so for instance they may both eat the same seeds. But this creates competition, seeds eaten by members of the first population are not available to the second, and vice versa. Now some small change in eating behavior happens in, say, one of the populations so that its members now start eating another type of seed. This behavioral change may result in selection for morphological differences (say, changes in beak size and shape), which results in what is termed *character displacement* (Pfennig and Pfennig 2012, Chapter 1).

There are two key features of this process that we wish to highlight. First, environmental overlap is a necessary condition for this process to operate. If the ranges of two populations do not overlap, then, to revert back to our example, seeds eaten by one population are not taken away from the other. Penguins and polar bears do not compete.

Second, phylogenetic closeness makes competition more likely. Why? For competition to occur between two populations they must be similar in some feature of their life histories. They must eat the same thing, or compete for the same mates, or whatever. So, while phylogenetic closeness is not a necessary condition for the ecological process of competition, it does make it much more likely.

When one population or species splits into two while maintaining environmental overlap (sympatric speciation), both of these conditions are present. The two are, by hypothesis, still overlapping, and are as phylogenetically close as can be.

In what follows we will present an argument at the species level. We do this not because we think the species level is *the* correct level for this analysis. Rather we think it is a good level for this sort of argument – ecologically, it is well established that competitive interactions do occur at the species level. Competition can also occur at subspecific levels (Pfennig and Pfennig 2012, Chapter 5), but it is harder to imagine it at much higher taxonomic levels. Also, we will here focus on the species level because we have one very useful fact about the total number of species, more precisely, the number of eukaryotic species.

How important is competition (leading to character displacement) in the diversification of species over evolutionary time? That is, when we look at diversification over some time period, how much of it is explained by competition? It turns out that this question can be answered, and the answer is "not much."

It is widely accepted that there are currently about 8 million eukaryotic species (Mora et al. 2011). It follows that there are 31,999,996,000,000 possible species pairs that could be drawn from this 8 million. Let's round up and call this 32 trillion pairs. Let us pick a pair: the little brown bat, *Myotis lucifugus,* and the sperm whale, *Physeter macrocephalus.* (We will here ignore what in other contexts would be an important point, namely, the groups bats and whales are now, and have historically been composed of many species. Here we focus on just two.)

So, of the 32 trillion pairs of currently extant eukaryotic species, how many of them are experiencing the ecological process of character displacement due to competition? An ecologically and phylogenetically plausible answer is 4 million. How do we get that? Each currently extant species is the result of a speciation event. Let us suppose every single one of these speciation events is sympatric (that is, one that keeps the daughter species together in the same environment) and results in the two species competing. If A competes with B, then by the definition of competition, B competes with A. But if A competes with B, then it is (probably) not competing with any C, because the dual conditions of environmental overlap and phylogenetic closeness will (likely) not be met. Thus 4 million competitive pairs out of the 8 million species. But that means that only 4 million of the 32 trillion possible pairs can possibly be subsumed under Darwin's hypothesis. That is, about 0.0000125 percent of all possible cases can possibly be so explained, or about one in 10 million. Which leaves a lot of divergence unexplained.

(Suppose our estimate of 4 million competitive pairs is off by a factor of 10, that is suppose that over its history every species has competitive interactions with 10 other species. That does not help matters much as it still leaves divergence in about 99.999999 percent of all species pairs unexplained.)

In contrast, the ZFEL is not an ecological hypothesis. It is the product of the sort of stochastic necessity of divergence between any two independently evolving lineages demonstrated in Section 4. Potentially it applies to all 32 trillion pairs of extant species. Of course, the ZFEL requires that the lineages be evolving independently of each other – an assumption that is explicitly violated in cases of competitive pairs, where their trajectories are causally linked, or coupled. Thus, to the extent that some pairs have trajectories linked by competition, the ZFEL does not apply. But, what we have just shown is that that number is very small compared to 32 trillion.

We do not need to develop any fancy philosophy of science account of goodness of explanation to say that a hypothesis that explains only 0.0000125 percent of its intended domain is not very good.

5.2.2 Hierarchical Selection-for-Difference

We have rejected the Darwinian principle of divergence as woefully inadequate to explain the diversification of life on Earth (or of life elsewhere, but we will ignore that for now). Before moving on to the ZFEL hypothesis we want to briefly cover two other possible responses to the question of diversification.

One way of summarizing the argument above is to say that the Darwinian competition hypothesis is much too limited to explain the broad pattern of diversification in nature, for instance that between bats and whales. Whales do not compete with bats and penguins do not compete with polar bears. But perhaps an explicitly hierarchical theory of competitive character displacement can rescue this idea.

Sister species compete with each other over restricted spatial and temporal scales. Genera also compete with each other over broader spatial and temporal scales. What about families and orders? Before going too far with this hypothesis we should address two different problems it faces. First, reproduction produces lineages. Lineages are real things, extending over space and time. Lineages split forming branching phylogenies. Branching phylogenies are hierarchical in that there are lineages nested within lineages. In particular, so long as reticulation (e.g., lateral gene transfer, or hybridization) is rare and so long as we don't try to apply the idea at or below the species level, monophyletic groups are formed by branching phylogenies. Monophyletic groups are real. Systematists name monophyletic groups. But do the ranks named across very different groups refer to some common reality. That is, is an order within chordates the same sort of thing as an order in green plants? If not, then we should be careful in using those ranks within a hierarchical selection theory.

The second problem is that phylogenetic entities are not necessarily ecological entities. Damuth (1985) first raised this sort of objection against the idea of species selection. Stanley (1975, 1979) thought that any sort of species sorting should count as species selection, ignoring the fact that chance alone will almost always result in some sort of differential survival and reproduction among species. A meaningful account of selection must be an ecological account that explains differential survival/reproduction in terms of differential fitness to an environment (Brandon 1990). Damuth argued that species are characterized genealogically and need not be, and usually aren't, ecological entities. Brandon (1990) and Eldredge (1985) both proposed that evolution requires dual hierarchies, one ecological and one genealogical. So, the hierarchical selection hypothesis under consideration would need to specify the relevant hierarchy of ecologically meaningful units, which may, or may not, closely correspond to the genealogical entities produced by the evolution of branching lineages.

This problem may not be insurmountable, but we will not try to resolve it as we think there is another closely related problem that is fatal. Selection seems less and less plausible as we increase in hierarchical rank (and this is so regardless of exactly how we specify the hierarchy). Although we are convinced that selection does operate at the level of groups of individuals, this is still controversial among biologists. We are even convinced that selection at least occasionally operates at higher levels. David Jablonski (1986) has presented strong evidence that during mass extinction events genus level selection has occurred. What he found was that genera with greater geographic ranges outsurvived genera with smaller geographic ranges, and that this effect could not be reduced to species geographic ranges. Genus geographic ranges *screened-off* (*sensu* Salmon 1971; also see Brandon 1990) species geographic ranges. Although the data do not directly support any particular ecological explanation of this fact, it seems reasonable that this is a form of bet-hedging. That is, during the hard times of a mass extinction genera with larger geographic ranges will more likely find a refuge and thus survive, than genera with smaller ranges. Thus, genus level selection occurs (perhaps not frequently, but still it exists).

Interestingly this is not the sort of selection that hierarchical selection-for-difference requires. Competition, which is required for selection-for-difference, means that the fitness of the two competitors are linked, such that what is fitter for one depends on the state of the other. This is importantly different from the case where in a certain environment a certain trait is favored in one group independent of what is happening in another. In the sense in which we have used the term, selection-for-difference is a second-order phenomenon, that is, it cannot be understood by looking at the evolution of lineages in isolation. In Jablonski's case, having a larger geographic range is good, regardless of what other lineages are doing.

So, to evaluate the hierarchical selection-for-difference hypothesis, we need to not just ask whether it is plausible that selection occurs at each of the relevant levels, we need to ask the more specific question of whether selection-for-difference is happening at these levels. This seems radically implausible. Bats are a higher taxon, and so are whales, and they have become very different from each other. But is their difference the result of some coupled ecological process that is pushing them farther and farther apart? How about two higher taxa with some degree of habitat overlap, like bears and whales? Is their difference the result of competition at the level of higher taxa?

For both questions, our answer is that we don't think so, nor are we aware of anyone who does. We have spent some time reviewing this hypothesis because it has the right logical form and the right spatial/temporal scale to explain the phenomena of diversification. But now we reject it.

5.2.3 Disparity as a Pseudo-problem

Next we consider a response to the problem of diversification that, again, no one has championed. But, again, we think it worthy of consideration. This response is to say that the problem of diversification is a pseudo-problem, that once we explain the trajectory of every evolutionary lineage, there is nothing left to explain. Consider the bat-whale case discussed earlier. The argument could be made that as bats evolve, they move toward an attractor, a fitness peak in morphospace that corresponds to a bat-like morphology. That is to say, over the tens of millions of years of bat evolution, "bat-ness" was favored by natural selection in this group. One could then say the same for whales, that as they evolve selection moves them toward a "whale-ness" attractor. And, the response continues, bats end up where they did because of the selective forces acting on bats and whales ended up where they did because of forces acting on whale, and that – the response concludes – is the end of the explanatory story. Put another way, once one has explained every lineage's trajectory there is nothing more to explain.

For anyone easily seduced by reductionism, this is appealing. But it is analogous to trying to explain the speed of a two-car draft at Talladega by looking at each car's speed in isolation. As we have seen, this will underestimate the speed of the two-car draft by 3–5 mph. This reductionist explanation is both empirically wrong and ontologically wrong – it misses the existence of, and causal relevance of, the two-car aerodynamic object. In what follows we offer a different account of what the reductionist with respect to diversification misses.

ZFEL-based diversification is a second-order process. Our argument has been that to explain a second-order process we must go beyond a mere concatenation of first-order explanations. Here we will use a notion of *explanatory robustness* to support this view. Doing so sheds further light on the phenomena of diversification. We will use two stories to illustrate when and why explanatory robustness is both apt and important.

Consider Adrian and Avery a happy couple who enjoy their time together. However, every March they separate for a weekend because Adrian is a philosopher and Avery is a psychologist. The third weekend of every March, the American Philosophical Association and the American Psychological Association meet, and every March, Adrian and Avery go to their respective APAs. One year the philosophers met in Chicago while the psychologists met in Cleveland. That weekend Adrian and Avery were 307 miles apart.

On the other hand, Blake and Brett were never a happy couple, their relationship was marked by violence and destruction until a judge stepped in and issued mutual restraining orders. GPS tracking devices were implanted behind their ears and a program installed on their phones that shows in real time both of their locations and the distance between them. According to the very serious judgment, if they get within 100 miles of each other both will go to jail. One weekend Brett was in Nashville and Blake in Asheville. According to the app on their phones they were 240 miles apart. Then Blake noticed Brett's icon moving east along I-40 toward Asheville. Blake waited an hour, then too worried to stay put started driving east on I-40, later taking I-77 north. The distance between Blake and Brett never got under 100 miles.

There are some issues in the theory of explanation raised by these examples that we think we can safely avoid. For instance, there is the question of whether all events are explainable (in principle). Some would have it that only highly probable events are explainable (e.g., Hempel 1965), while others (e.g., Salmon 1971, 1984) argue that highly probable and highly improbable events are equally explainable. We side with Salmon, but for our current point nothing hangs on this. For the purposes of this discussion we will suppose that all events are explainable, but that some events have deeper, more robust, explanations than others. In science the notion of robustness of a result in a model is widely used. A robust result in a model is one that happens over a wide range of parameter values, or a wide range of initial conditions. Similarly, we can say that a robust event in the world is one that would have happened over a wide range of initial conditions. Robustness in this sense is the flip side of contingency. The notion of contingency at work here is that promoted by Gould (see esp. Gould 1989) and is familiar to students of human history (see Fleming and Brandon 2015; Sterelny 2016). World War I was highly a highly contingent event – had Archduke Ferdinand's driver not gotten lost in Sarajevo or had Gavrilo Princip, who was a poor marksman shot more in accordance with his abilities – World War I would not have happened. But it did, and arguably, after the Treaty of Versailles, World War II was nearly inevitable. It would be a misunderstanding of history if we failed to make this distinction between World Wars I and II.

Back to our two couples. We can robustly explain why Adrian and Avery separate the third weekend of every March. That result is robust. Given their dedication to their respective professions there is little that would keep this from happening. In contrast, the fact that one weekend they happened to be 307 miles apart has, at best, a very shallow, or very fragile explanation. While we have a deep understanding of why they go to their meetings, the distance between them that results is incidental and will vary from year to year. The distance between Brett and Blake, or more precisely the minimum distance between them, has a robust explanation based on the nature of the court order, and their motivation to stay out of jail.

Let us now return to the standard explanation of macroevolutionary disparity. Let us suppose that the trajectories of bat and whale evolution were driven by ecology, with each group moving toward its own adaptive "attractor," the lineage leading to bats moving to the bat-ness area of morphospace and the lineage leading to whales moving to the whale-ness area. This is then a case of different lineages evolving randomly with respect to each other, with the result being increasing disparity. But Darwin's stab at this and contemporary biology's use of this explanation touches only half of the process. It leaves out the second-order stochastic process detailed in Section 4 and is inadequate in principle. Second-order phenomena require second-order explanations.

The stories of our two couples make clear just what is missing from the standard explanation of diversification, the bat-attractor and whale-attractor stories. That first-order explanation leaves the distance between bats and whales on a par with the distance between Avery and Adrian when they were 307 miles apart. That is a fragile result. But we have two good reasons to think that increasing disparity is a robust result, like the minimum distance between Blake and Brett or the fact that Adrian and Avery are likely to end up apart every year at conference time. First, there is the theory we have developed previously (McShea and Brandon 2010) and here that shows that increasing disparity is a probabilistically expected result under a wide (and realistic) set of conditions. Second there is the evidence coming from the history of life on Earth, in particular the history of diversity/disparity increases after every mass extinction (see McShea and Brandon 2010, Chapter 3). This evidence is important in countering the charge that life started at minimum diversity/disparity and had no way to go but up. But after mass extinctions, going down is an option, but an option not taken. Thus, for both theoretical and empirical reasons we think the disparity increase is a robust phenomenon that requires a robust explanation. The independent-attractor explanation is not robust and so does not in fact adequately explain the phenomenon.

This argument does not apply to either of the selection-for-difference accounts we have discussed above. Both of those accounts correctly treat disparity increases as a robust phenomenon. The problems with those accounts are not that they are of the wrong kind, but rather that seem to explain very little of the domain for which we want an account.

That leaves us with the ZFEL account. It is of the right sort – it treats disparity increase as a robust phenomenon requiring robust explanation. It is logically and mathematically correct. And its assumptions are pretty minimal – it just requires independence between evolving lineages. That is a condition that surely is often met: consider our favorite example of bats and whales. Or pick a random pair of higher taxa. Thus, for a large majority of cases, the ZFEL is the best available explanation of macroevolutionary diversity/disparity.

5.3 Complexity

Diversity and complexity have had very different upbringings in biology over the past 150 years since Darwin. Diversity has always been a respectable child, welcome in polite scientific circles. There has been general agreement on what the term means and how to measure it, and it has attracted a great deal of focused and serious research. The complexity of organisms, however, is the evil twin. The concept was treated badly from the start, when Darwin – so clear thinking on so many things – appeared to conflate it with notions like organization, degree of development, and adaptedness (Darwin, Notebook E). The abuse continues today, as popularizations by professionals in biology, science journalists, and even popular culture invoke complexity whenever certain transparently unscientific notions needed to be discussed – the supreme superiority of humans, the Great Chain of Being, and evolutionary progress. All of these are considered off limits, in many biological circles, because they are value laden and science is supposed to be value free. But complexity *sounds* value free, so we can safely call humans complex. And we can safely call the evolutionary "advancement" from microbe to worm to fish to human a rise in complexity. When the word complexity is used, it sounds like we know what we are talking about, and – in terms the Victorians would have used – appearances are preserved.

Given complexity's history of bad treatment, it is no wonder that it has been so troublesome, and that discussion of it has been largely kept out of serious scientific journals (Ruse 2009), at least in evolutionary biology. Nor is it any wonder that the study of complexity has been so theory and data poor.

5.3.1 Theory

The discussion in the literature on the causes of complexity change is thin, consisting – with a few exceptions – of casual remarks offered *en passant*, a few short and plausible stories about the factors that might affect the evolution of complexity. In what follows, we call them all hypotheses, although some of them barely qualify as that.

Two main hypotheses that have been proposed, both explain why increasing complexity ought to be the expectation. One is the differentiation of repeated parts, a notion that originated in early twentieth-century paleontology (Williston 1914; Gregory 1935). The suggestion was that in a series of repeated parts, different parts would be under selection for different functions, and would therefore tend to differentiate. For example, in early terrestrial vertebrates, anterior vertebrae were selected for a new function, head support, in addition to the older function of transmitting swimming forces. As a result, anterior and posterior vertebrae became more different, increasing column complexity. In our terms, we want to say that to the extent that selection for head support and for swimming were independent, anterior and posterior columns can be said to be changing randomly with respect to each other, at least in the initial stages of differentiation. And therefore, this is at least partly the ZFEL. For clarity, let us add that it is both the fact of divergence and its magnitude that is explained by the ZFEL, but not any particular sequence of morphological changes in the vertebral column.

Modern biology invokes this same process. The standard explanation for the duplication and selective divergence of genes is the ZFEL (McShea and Brandon 2010), although not called that.

The other main hypothesis originated with Darwin, who suggested that the complexity of organisms might be favored by selection on account of the advantages of greater division of labor among parts. This is not the ZFEL, because here selection directly favors difference. For example, suppose that the greater differentiation of the mammalian tooth row, compared to the reptilian one, was the result of selection for more efficient food processing. In other words, the "labor" of food processing became divided, say, between one set of teeth specialized for piercing or slicing and another set specialized for crushing. The complexity of the tooth row increased, but to the extent that the teeth were selected together, to complement each other and cooperate in the performance of a single function, they would not have been changing randomly with respect to each other. Rather, they would have been selected *for* being different from each other. Or consider the asymmetrically positioned ear openings in some owls, which presumably arose by selection for better sound localization. If so, then this was selection for differentiation of ear location rather than independent selection on each ear, and therefore not the ZFEL. In general, selection for what today is sometimes called subfunctionalization, the splitting of a task among parts is not the ZFEL.

Another hypothesis is that there is kind of complexity ratchet. Decades ago, Saunders and Ho (1976) proposed that complexity would tend to increase in evolution because existing parts tend to become integrated into development and therefore difficult to remove, while new parts could be added without disrupting present function. The idea has been developed more fully recently under the heading of "constructive neutral evolution," or CNE (Stoltzfus 1999; Lukeš et al. 2011; Brunet and Doolittle 2018). In CNE, dependencies arise between initially useless parts and functional ones such that the useless ones become locked in, and their removal is selected against. CNE is ZFEL-like in that it invokes the random addition of new part types, but differs in that it requires selection to block reversals. Another ratchet, proposed by Lynch and Conery (2000; also Lynch and Force 2000), occurs when neutral additions in DNA go to fixation in small populations. Despite their affinities to the ZFEL, neither of these is a pure ZFEL process. The ZFEL has no requirement for negative selection or for small population size. The ZFEL also does not require that new parts have a neutral origin. New part types can arise when existing parts are selected for new functions (as in the vertebral-column example above), that is, when they change randomly with respect to other parts.

Other selection-driven mechanisms for complexity increase have been proposed. Waddington (1969) raised the possibility that greater complexity means greater ecological specialization, which in turn might be generally favored by selection. As species diversity increases, niches become more complex (because niches are partly defined by existing species). The more complex niches are then filled by more complex organisms, which further increases niche complexity, and so on (see also Darwin 1987). Bonner (1988, 2004) argued that complexity increases with body size, because large organisms require more specialized structures and therefore more parts. There are more such possibilities (McShea and Brandon 2010; McShea 1991, 1996), all interesting and some testable. None are the ZFEL.

It would be natural – but wrong – to think that in explaining a rise in complexity, the ZFEL and the causes listed above are alternatives. It is wrong, because the ZFEL is always at work. The question raised by these hypotheses is whether selection has acted in addition, augmenting the tendency to increase arising from the ZFEL. Recall from Section 3 that what the ZFEL offers is more than a hypothesis about the cause of complexity change, it is a general scheme for understanding change. It says that non-ZFEL selective causes need to be looked at in the context of the ZFEL. Just as we need to decompose the motion of the ball being tossed on a train into a component due to the train and another due to the toss, we need to decompose actual change in complexity into a component due to the ZFEL and another due to forces like those listed above. In other words, to understand and to measure the effect of those forces, we need to first green-screen the ZFEL expectation, so to speak, into the background.

Most of the chat in the literature about complexity has focused on gains. But there have also been remarks about complexity loss and the retention of simplicity (O'Malley et al. 2016, McShea and Brandon 2010). Commonly mentioned is the supposed simplicity of most parasites, under selection for loss of various functions, and therefore for loss of the parts that perform those function. Simplification presumably occurs when a parasite can take advantage of the fact that the host is performing these functions for it. (A similar argument explains the "complexity drain," mentioned in Section 2 and discussed further in Section 5.5.) Another selection-driven route to loss is the reduction in size that accompanied the move of the so-called meiofauna to interstitial environments, to life amid the sand grains of terrestrial and aquatic sediments. Selection for reduction in size would inevitably have been accompanied by loss of large part types, that is, many tissues and organs. A third route to loss is the stabilizing selection presumed to explain the continuation of simple organisms in simple niches. Hardly ever mentioned but hovering in the background of any discussion of complexity trends is the assumption that organisms with few part types – among multicellulars, say, sponges – that are thriving at a low complexity, would experience intense stabilizing selection to retain their present simple form. We discuss this further in a moment.

As for gains in complexity, losses too can only be understood against the backdrop of the ZFEL. But there is an asymmetry here. When complexity increases, it could be the result of the ZFEL alone or the ZFEL plus some selective force favoring gain. However, when complexity decreases, it can only be the result of selective opposition to the ZFEL. For both, the great virtue of a quantified ZFEL, of course, is that it gives us a way to quantitatively parse any observed complexity increase, decrease, or stasis into its ZFEL and selective components.

5.3.2 Data

The data on organismal complexity consists first of a handful of studies on interesting correlates. Marcus (2005) found a correlation between complexity and gene number in bacteria. Bell and Mooers demonstrated a correlation between complexity and body size (see also Bonner 1988, 2004). And Schopf et al. (1975) found an interesting *positive* correlation between complexity and extinction probability. (The correlation could be spurious, they said, but if it's not, complexity would seem to be a bad thing.) There has also been some empirical work on trends in complexity, demonstrating for example an increasing trend in vertebral column complexity in the transitions from fish to reptile to mammal (McShea 1993; Buchholtz and Wolkovich 2005), a decreasing trend in number of types of bony elements in the skulls of mammal-like reptiles (Sidor 2001), and an increasing trend in number of limb-pair types in arthropods (Adamowicz et al. 2008). Most trend studies have been quite narrow, focused on a single phylum. The exception – to which we now turn – is a study two decades ago on number of cell types.

The Valentine et al. Study. The landmark dataset on diversity is Jack Sepkoski's curve (Sepkoski 1978), showing changes in number of marine-animal families over the Phanerozoic Era, the last 540 million years. Sadly, complexity has nothing like that, no curve with so many data points at such a high level of temporal and taxonomic resolution. But it does have one data set with the same temporal scope. Figure 9A shows Valentine et al.'s (1994) graph of the Phanerozoic trajectory of maximum number of cell types in animals. We discuss this curve at some length, partly because it offers a window into the history of complexity, a picture window through which we can take in a fair bit of the complexity landscape. But also because it provides a nice way to explore the ZFEL and the dynamics underlying complexity trends.

In Valentine's graph, number of cell types in certain animals is plotted against the time of origin of their groups. All cell-type counts were based on living organisms in modern groups (since cell types cannot ordinarily be distinguished in fossils), and the assumption was that the earliest members of these groups had roughly the same number of cell types as the moderns. For example, they assumed that the earliest arthropods had about the same

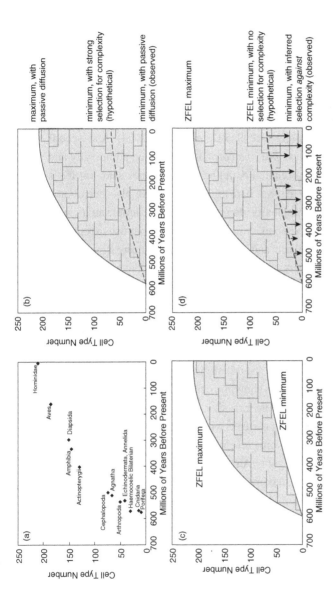

Figure 9 (A) The original Valentine data, showing the rise in the maximum for number of cell types over the history of the Metazoa. (The timescale is out of date, but using an updated timescale does not substantially change the shape of the curve.) (B) The standard null model, diffusion from a lower bound. The graph shows the observed pattern as a region (shaded area) defined by the upward trajectory of the maximum and a constant minimum. In the standard null model, the expectation is that taxa like sponges with fewer cell types will be present throughout the history of the group. The graph also shows the hypothetical rise in the minimum that, according to the standard model, would be expected if selection strongly favored complexity (dashed curve) and the simplest organisms were eliminated. Clearly the observed pattern does not match the pattern expected with strong selection. (C) The ZFEL null model. The shaded area shows the expectation using the ZFEL null if the ZFEL were acting alone, without selection. The maximum rises, but so does the minimum, both due to the ZFEL. In the absence of selection, all lineages have a tendency to rise in complexity. (D) The observed pattern again, showing the rise of the minimum predicted by the ZFEL (dashed line) with arrows showing selective forces that are inferred to be present to bring down the minimum, to match what is observed.

number of cell types as modern arthropods (about 50). Notice that the data collected in this way could be misleading, say, if there were many wholly extinct groups with large numbers of cell types or if cell-type counts within any of the modern groups have significantly declined over the history of the group. But taking the data at face value, they reveal a steady rise in the complexity of animals. More precisely, they show a rise in the maximum, in the complexity of the most complex animal in existence at given time, an expansion of the upper envelope of complexity. An important caveat: the data show only what is happening at the level of cell types. Recall that on account of the level relativity of complexity, part counts at the level of, say, organs or even molecules could show a very different pattern, perhaps a decreasing trend or no trend at all.

In addition to the rise in the maximum, possibly the most important feature of Figure 9A is one that is not marked explicitly. And that is the stability of the minimum. The modern animals with the fewest number of cell types are sponges (Porifera) with about 10, and as required by their method, Valentine et al. assume the earliest sponges had about the same number. Let us grant that this is the case. Now since sponges have been present continuously over the Phanerozoic, it follows that the minimum number of cell types for animals over that whole time, has remained pretty constant, at about 10. It turns out that these two features of the graph – a rising maximum and a constant minimum – can tell us something about the forces governing change in complexity. However, what they tell us depends critically on our choice of null model.

The Standard Null: Diffusion. Is the trend in Figure 9A the result of selection *for* complexity, perhaps arising from the advantages of dividing labor? To answer this, to understand what Figure 9A is telling us, we need a null model. To know if selection has been acting, we have to know what would have happened in its absence.

Figure 9B shows the standard null model that has been used for macroevolutionary trends generally. The model has been used to look at trends in a variety of variables, especially body size but also complexity. Here, for heuristic purposes, we have layered it onto the axes of the Valentine graph. In the model, the first animals are assumed to be simple, having few cell types, and complexity changes – increasing or decreasing – only at speciation events, shown as right-angle up and down branches from existing species (horizontal lines). In this null model, increases and decrease are equally probable. In other words, there is no selection so there is no bias in the direction of change, no upward tendency. Nevertheless, the mean rises, and the reason is that the expansion of the group is limited below by a boundary, a lower limit on the number of cell types an animal can have. In principle, that lower limit could be 1 – the ancestral protist condition, a single cell type – but more likely the lower limit for animals was somewhat higher, perhaps close to 10, the number of cell types in a sponge. The process has also been described as bounded diffusion (McShea 1994; Gould 1996).

Against this null, what is the expectation for the effect of selection favoring increases? The maximum should rise, and it does. But of course, this would be expected even in the absence of selection. And the minimum should rise also. Simple animals like sponges would be subject to strong selection for increase, and – constraints permitting – would increase in complexity. And at some point in their ascent up the complexity scale, their form would change, and they would no longer be sponges. But apparently sponges have persisted over the Phanerozoic, and the minimum has remained roughly stable. In other words, on account of the stable minimum, the pattern does not fit expectation for complexity under strong selection, but it does fit the standard null model.

This is the null model that one of us used in a study of vertebral column complexity (McShea 1993). It is the null model that Valentine used. And it is the null model that Stephen Jay Gould (1996) used in his analysis of the putative trend in complexity in his book *Full House*. Gould was writing about the rise over the history of life of what he called "excellence," but he also used the word complexity, albeit mostly in the colloquial sense.

The ZFEL Null. The use of some null model is certainly appropriate, and logic of the analysis based on the behavior of maximum, mean, and minimum is right. But for complexity, the standard null is the wrong one. As the ZFEL tells us, the expectation for complexity in the absence of selection is not a 50–50 probability of increase and decrease. It is a greater probability of increase in all lineages. The great majority of the variations arising is a set of parts will tend to make them more different from each other. In the absence of selection, cell types should spontaneously differentiate. Thus, the appropriate null model is one in which simple animals do not stay simple very long, and the minimum rises. This is the ZFEL null, shown in Figure 9C.

Notice this reasoning does not apply to first-order variables like body size (Stanley 1973; Gould 1988; McShea 1994). For these, the standard null is the right one. It is second-order variables like diversity and complexity, measures of variance, that need an upward-biased, ZFEL-based null.

Based on a ZFEL null, the interpretation of the Valentine curve changes. Figure 9D shows the Valentine curve with the ZFEL expectation of a rising minimum overlaid. Against this expectation, it is clear that since the minimum did not rise, some force, presumably selection or perhaps constraints, acted to keep the minimum low. If it was selection, then it would have been selection for complexity stasis, for keeping sponges and others at the same number of cell types at which they originated. And in the context of the ZFEL, which imparts a constant tendency for complexity to increase, selection for stasis amounts to selection *against* complexity.

If it were possible to insert a pause in a written narrative, we would insert one here, to let that last conclusion sink in. Our intellectual reflexes in evolutionary biology tell us that complexity is generally a good thing, favored by selection. Perhaps this is because we are so used to thinking of humans as complex, as an endpoint of a pervasive complexity trend across all animals, if not all life, a trend driven by natural selection. And natural selection favors good things. Whatever the merits of this line of thought, and they are few, it can be relevant only to colloquial complexity, whatever that is. For complexity at the level of cells in the Metazoa, the opposite seems to be true. The prevailing winds have been blowing against complexity.

Let us put it another way. Turned loose in a classroom, the young children run riot, while the teacher struggles to restrain them, to bring order. Left alone, cells would spontaneously differentiate, raising complexity, while selection struggles to bring order, to limit the increases in complexity to the tiny subset that works. A recent review of our first book by Vlieger (2019) expressed the point nicely: "change is the constant, with natural selection acting … to rein in the creative maelstrom generated by the ZFEL."

One final point to make here. The change in interpretation that arises simply from adopting the right null model is dramatic. Using the standard null, which we argue is inappropriate for second-order variables like complexity and diversity, the observed data are consistent with the null, and no forces need to be invoked. Using the ZFEL null, which is the right one here, the observed data are inconsistent with that null, and a strong downward force is inferred to be present.

5.3.3 Hierarchy, the ZFEL, and an Evolutionary Syndrome

Hierarchy Is Not Complexity. As we have said, diversity's evil twin, complexity, is a trouble-maker on account its constant conflation with colloquial complexity. But the twin's troubles are actually worse than that. It is also afflicted with a multiple personality disorder. The same word, complexity, is routinely used not only for organisms with many part types, but for those with multiple levels of organization. This second usage has to do with hierarchy, levels of nestedness of parts within wholes. If a single bacterial cell occupies the bottom hierarchical level, then a eukaryotic cell occupies the second level, since it arose historically as an association of bacterial cells. A eukaryotic multicellular individual occupies the third level, since it consists of an association – a clone, in most cases – of eukaryotic cells. And a colony, like a bryozoan colony or a baboon troop, therefore occupies a fourth level. Complexity in this vertical sense, number of levels, has zero conceptual overlap with complexity in the horizontal sense, number of part types. They are conceptually independent. In principle, an organism can be hierarchically deep and part-type poor, or hierarchically shallow and part-type rich, or either of the remaining two combinations. Furthermore, part-type complexity is a level-relative concept, and hierarchy is not. (How could a count of levels be level relative?) Despite this nonoverlap, such is the slipperiness of language that it is easy to imagine ourselves saying in a thoughtless moment that human society is complex because of the many different social roles we have (part-type complexity), made possible maybe by our sophisticated use of language, and then concluding that we are hierarchically more complex – that we occupy a "higher level of organization" – than a baboon troop. Language was a hugely important innovation for people, and we may in fact be hierarchically higher than a baboon troop, by some understanding of hierarchy, but the complexity of our social lives is something altogether different. Worse than apples and oranges, part-type complexity and hierarchy are aardvarks and orchids, not even in the same conceptual kingdom.

As mentioned in Section 2, here we avoid any possibility of confusion by using the word complexity only to refer to number of part types and the word hierarchy for number of levels.

We dwell on this separation between complexity and hierarchy mainly to explain that the ZFEL has nothing directly to say about hierarchy. The ZFEL tells us to expect number of part types to increase at any given level. By itself, it does not tell us to expect number of levels to increase. It does not predict the trend in hierarchy documented by McShea (2001), showing the trajectory of increase in number of levels over the history of life. Nor does the ZFEL predict Maynard Smith and Szathmáry's (1995) "major transitions," a motley collection of episodes in evolution, many (but not all) of which involved a rise in hierarchy.

The ZFEL in Action. All of this so far has been a conceptual critique. Nothing has been said about empirical connections between complexity and hierarchy. Even though they are conceptually independent, there could still be causal connections between them. For example, a systems-theoretical argument has been devised for why we would expect a new hierarchical level $N + 1$ to arise whenever a critical threshold in number of part types is crossed at level N (Fleming 2012). That is just one of the many possible avenues that we could explore. Here we focus on one that illustrates not just the obvious role that the ZFEL has played in the differentiation of parts, like the increase in cell types in animals, but the role that such differentiation might play in a pattern at an even bigger scale.

More precisely, we will look at three patterns, three trends in evolution: (1) the association of individuals to form new higher-level wholes (that is the rise in hierarchy); (2) the increase in number of part types within these wholes (as illustrated by the Valentine data); and (3) the complexity drain, or the loss of parts within these parts, in other words, a loss of subparts, two levels down from the whole. So for example, as a multicellular individual arose from ancient clones of protists (trend 1), the cells differentiated to form an individual with many cell types (trend 2), and these cells in turn lost many of their internal parts, becoming in the end simpler on average than the protists they evolved from (trend 3). There is data to support this last step. Cells in multicellular individuals have fewer part types, on average, than free-living protists (McShea 2002). To pick an extreme contrast, an *Amoeba* has many, many part types, while human blood cells have no macroscale parts at all. In the language that one of us has used to describe this process, in the transition to multicellularity, cells become "machinified." They are transformed from complex, omnicompetent, supple organisms into simple, specialized, and somewhat brittle little machines (McShea 2015).

The factors driving these three trends doubtless include both the ZFEL and selection. One story might be that new levels arise, perhaps on account of selection for larger body size (Bonner 1988, 2004). The parts then differentiate, driven by the ZFEL, and perhaps under selection for division . of labor as well (Bonner 1988, 2004). Finally, the complexity drain within these parts is driven by selection. As the multicellular individual emerges, it takes over many of the functions formerly performed by the individual cells, and selection favors the loss of subcellular parts that performed those functions. As for the putative reduced complexity of parasites, the driving force is selection for streamlining.

The possible large-scale pattern is the repetition of these trends across levels. The bacterial symbionts that joined to form the first eukaryotic cell were already different, but they became more differentiated, losing some of their internal parts as they did so. And at a much higher level, in the evolutionary origin of colonies and societies, the proposal is that the multicellular colony members become more differentiated and at the same time less internally complex (McShea 2002).

The causal story could be roughly as outlined above. Or it could that there is a different causal story at every level, or perhaps in every single instance. Or there could be a single cause, an unknown factor that causes all three trends at every level. Until we know, it may be best not to stake out a position, and instead to think of the three trends as a kind of syndrome, a combination of "symptoms" associated with the rise of hierarchy, perhaps with a single underlying – but still unknown – cause (McShea 2015).

In any case, it is clear that the ZFEL has two roles here, both central. The obvious role is in the production of new part types. For multicellulars, that is the increase in cell types. But the ZFEL also has an important background role in the complexity drain. It is easy to say that cells are under selection for stream-lining but consider how difficult that must be to accomplish, given that the ZFEL is at every evolutionary moment pressing toward *greater* differentiation, toward *more* part types. Loss of any given part-type may or may not be easy, but either way a net overall loss of part types is quite likely very difficult. It is an evolutionary struggle, against the headwind of the ZFEL. As our bumper sticker–length summary of the ZFEL says, complexity is easy, simplicity is hard (McShea and Brandon 2010).

There are any number of way that a quantified ZFEL could be useful here. We will suggest just one. It would allow us to measure the intensity of the forces driving differentiation – presumably the ZFEL plus selection for division of labor – and also to measure the intensity with which the ZFEL is opposed in the complexity drain. If the two processes are in fact linked, as the syndrome suggests, the expectation is that the intensities of those forces will be correlated. A quantified ZFEL enables us to test.

5.4 Some Educated Guesses

Section 4 was empiricist in its orientation. Let us not guess at what factors were responsible in this instance of diversification or in that instance of complexification. Let us instead gather data so that we can rigorously test the ZFEL null model. If we fail to reject it in a particular case, then further investigation may well lead us to accept the ZFEL as responsible. Let's be careful and modest and examine things in a quantitatively rigorous way before drawing conclusions.

In this short section we throw epistemological caution to the wind and answer the question: what are our best guesses about what is responsible for the large-scale patterns we see in evolution for complexity and diversity? For the rise in complexity, there is a generalization that seems fairly well supported. It is that the ZFEL is *not* the whole story. Similarly, with respect to the rise of diversity (broadly understood, i.e., disparity) in the history of life there is a generalization that seems hard to dismiss. It is that the ZFEL *is* pretty much the whole story.

Since we have shown that diversity and complexity are really the same thing, how can it be that we draw different conclusions about them? The answer has to do with the fact that the term *complexity* is largely used at one hierarchical level, namely, the organismic. Sometimes we talk about the complexity of cells and of societies, but the organismic level dominates usage within biology. In contrast, although we can sensibly talk about the diversity of a group of cells, diversity is most commonly applied at the species and higher levels. Complexity and diversity are one and the same thing, but standard usage tends to apply them at different hierarchical levels. As covered in Section 5.3.2, complexity at the organismic level has pretty consistently been selected against. The ZFEL produces organismic complexity and selection knocks it down, because complexity, in general, is not a good thing. For perhaps romantic reasons, biologists have tended to view complexity as a marvel that requires selection (and before that, God) to explain. Any good engineer would be skeptical. As the brilliant twentieth-century automotive engineer Colin Chapman said in explaining his racing philosophy, "Simplify, then add lightness."[3]

But for diversity at and above the species level the argument given in Section 5.2 is that the ZFEL is the only plausible explanation for the vast majority (say, about 99.9999 percent) of cases of diversification. No other available explanation works. Of course, we are more than happy to see rigorous investigations of this or that case, but the very general arguments we have gone over make this conclusion hard to resist.

How can this be? The answer, we think, is that organisms are targets of selection and when they perform badly selection ruthlessly eliminates them. Species and higher-level clades may well be targets of selection occasionally (recall Jablonski's data on clade-level selection during mass extinctions), but at present, it seems reasonable to think that this higher-level selection is less pervasive, less powerful than organismic selection. Butterflies are very diverse as a group, and beetles even more so, but there does not seem to be a functional entity, butterflies or beetles, that is affected one way or another by that diversity. Or to take our favorite example, whales and bats have radically diverged since their last common ancestor. But there is no bat-whale functional unit that has to answer to some high-level selection pressure for simplicity. Radical and beautiful and awe-inspiring diversity is cost-free at higher hierarchical levels.

---

[3] www.lotuscars.com/about-us/lotus-philosophy.

5.5 Coda

Heraclitus was right. You can't step in the same river twice. In his terms, everything flows. In ours, everything varies. Biology has known this at least since Darwin. Physics and chemistry, the so-called hard sciences, have been slower to learn it. This simple principle – everything varies – combined with heredity, in systems with multiple individuals or multiple parts – leads to a powerful conclusion: diversity and complexity tend to increase. *And that is the foundation for the missing two-thirds of evolutionary theory.*

Biology has long understood the underlying pieces of this argument, embracing the principle of heredity and allowing spontaneous variation where necessary. But it proclaims to be baffled by complexity, and to a lesser extent by diversity. What explains the miracles of brains and rain forests? From the standpoint of the ZFEL, this is a strange worry – it is a worry about the existence of the stochastically inevitable. Now, to be clear, it is a worthy puzzle why diversity took the form that it did, why we have on this planet the multinucleate *Paramecium*, the creeping *Dictyostelium*, the towering *Sequoia*, and the behemoth *Balaenoptera*. And it is a worthy puzzle why complexity took the form that it did, why we have individuals with limbs, larynxes and lymphocytes. Worthy because there is no principle that explains why we have *these* taxa, with *these* parts. But it should be no mystery that there are *many* taxa, and that some of them have *many* parts. Yet complexity remains one of biology's most widely acknowledged – and yet at the same time, least studied – mysteries. Diversity has been more studied and is better understood, but the principle that constitutes its first cause – what we call the ZFEL – has been left implicit, unstated.

As we said at the start, variation and heredity together do not guarantee we will get diversity and complexity. We did not get it on Mars. We got it, but it did not last on the Yucatan Peninsula 65 million years ago, ground zero for the Cretaceous-Tertiary impact. And in some places – the paved-over parking lots around the world – we have recently lost it. But even in these places, the universe tilts toward diversity and complexity.

Bibliography

Adamowicz, S. J., A. Purvis, and M. A. Wills. 2008. Increasing morphological complexity in multiple parallel lineages of the Crustacea. *Proceedings of the National Academy of Sciences of the United States of America* **105**: 4786–91.

Bell, G., and A. O. Mooers. 1997. Size and complexity among multicellular organisms. *Biological Journal of the Linnean Society* **60**: 345–63.

Bonner, J. T. 1988. *The Evolution of Complexity by Means of Natural Selection.* Princeton University Press, Princeton.

Bonner, J. T. 2004. Perspective: The size-complexity rule. *Evolution* **58**:1883–90.

Bookstein, F. L. 1988. Random walk and the biometrics of morphological characters. *Evolutionary Biology* **23**: 369–98.

Brandon, R. N. 1990. *Adaptation and Environment.* Princeton University Press, Princeton.

Brandon, R. N. 2006. The principle of drift: Biology's first law. *Journal of Philosophy* **103**: 319–35.

Brandon, R. N., and J. Antonovics. 1996. The coevolution of organism and environment. Pp. 161–78 in R. N. Brandon (ed.), *Concepts and Methods in Evolutionary Biology.* Cambridge University Press, Cambridge.

Bromham, L. 2011. Wandering drunks and general lawlessness in biology: Does diversity and complexity tend to increase in evolutionary systems? *Biology and Philosophy* **26**: 915–33.

Brunet, T. D. P., and W. F. Doolittle. 2018. The generality of constructive neutral evolution. *Biology and Philosophy* **33**: 1–25.

Buchholtz, E. A., and E. H. Wolkovich. 2005. Vertebral osteology and complexity in *Lagenorhynchus acutus*. *Marine Mammal Science* **21**: 411–28.

Ciampaglio, C. N., M. Kemp, and D. W. McShea. 2001. Detecting changes in morphospace occupation patterns in the fossil record: Characterization and analysis of measures of disparity. *Paleobiology* **27**: 695–715.

Damuth, J. 1985. Selection among "species": A formulation in terms of natural functional units. *Evolution* **39**: 1132–46.

Darwin, C. 1859. *On the Origin of Species.* John Murray, London.

Darwin, C. 1987. Notebook E. Charles Darwin's Notebooks, 1836–1844. P. H. Barrett, P. J. Gautrey, S. Herbert, D. Kohn, and S. Smith (Eds.), *Geology, Transmutation of Species, Metaphysical Enquiries.* Cornell University Press, Ithaca.

Deline, B., J.M. Greenwood, J.W. Clark, M.N. Puttick, K.J. Peterson, and P.C.J. Donoghue. 2018. Evolution of metazoan morphological disparity. *Proc Natl Acad Sci USA* 15(38): E8909–E8918.

Doolittle, W. F. 2012. A ratchet for protein complexity. *Nature* **481**: 270–71.

Drury, J. P., G. F. Grether, T. Garland Jr., and H. Morlon. 2018. An assessment of phylogenetic tools for analyzing the interplay between interspecific interactions and phenotypic evolution. *Systematic Biology* **67**: 413–27.

Eldredge, N. 1985. *The Unfinished Synthesis*. Oxford University Press, Oxford.

Finnegan, G. C., V. Hanson-Smith, T. H. Stevens, and J. W. Thornton. 2012. Evolution of increased complexity in a molecular machine. *Nature* **481**: 360–64.

Fleming, L. 2012. Network theory and the formation of groups without evolutionary forces. *Evolutionary Biology* **39**(1): 94–105.

Fleming, L. 2013. The notion of limited perfect adaptedness in Darwin's principle of divergence. *Perspectives on Science* **21**: 1–22.

Fleming, L., and R. Brandon 2015. Why flying dogs are rare: A general theory of luck in evolutionary transitions. *Studies in History and Philosophy of Science, Part C* **49**: 24–31.

Foote, M. 1994. Morphological disparity in Ordovician-Devonian crinoids and the early saturation of morphological space. *Paleobiology* **20**: 320–44.

Gingerich, P. D. 1993. Quantification and comparison of evolutionary rates. *American Journal of Science* **293**(A): 453–78.

Goodman, N. 1955. *Fact, Fiction and Forecast*. Harvard University Press, Cambridge, MA.

Gould, S. J. 1988. Trends as changes in variance: A new slant on progress and directionality in evolution. *Journal of Paleontology* **62**: 319–29.

Gould, S. J. 1989. *Wonderful Life: The Burgess Shale and the Nature of History*. W. W. Norton, New York.

Gould, S. J. 1996. *Full House: The Spread of Excellence from Plato to Darwin*. Harmony Books, New York.

Gould, S. J., and R. C. Lewontin. 1979. The spandrels of San Marco and the panglossian paradigm: A critique of the adaptationist programme. *Proceedings of the Royal Society of London, Series B* **205**: 581–98.

Grant, P. R., and B. R. Grant. 2006. Evolution of character displacement in Darwin's finches. *Science* **313**: 224–26.

Grant, P. R., and B. R. Grant. 2008. *How and Why Species Multiply: The Radiation of Darwin's Finches*. Princeton University Press, Princeton, NJ.

Gregory, W. K. 1935. Reduplication in evolution. *Quarterly Review of Biology* **10**: 272–90.

Hansen, T. F., and E. P. Martins. 1996. Translating between microevolutionary process and macroevolutionary patterns: The correlation structure of inter-specific data. *Evolution* **50**: 1404–17.

Hempel, C. 1965. *Aspects of Scientific Explanation*. The Free Press, New York.

Hunt, G. 2006. Fitting and comparing models of phyletic evolution: Random walks and beyond. *Paleobiology* **32**: 578–601.

Hunt, G. 2007. The relative importance of directional change, random walks, and stasis in the evolution of fossil lineages. *Proceedings of the National Academy of Sciences of the United States of America* **104**: 18404–8.

Hunt, G., S. A. Wicaksono, J. E. Brown, and K. Macleod. 2010. Climate-driven body-size trends in the ostracod fauna of the deep Indian Ocean. *Palaeontology* **53**: 1255–68.

Jablonski, D. 1986. Background and mass extinctions: The alternation of macroevolutionary regimes. *Science* **231**: 129–33.

Lande, R. 1976. Natural selection and random genetic drift in phenotypic evolution. *Evolution* **30**: 314–34.

Lukeš, J., J. M. Archibald, P. J. Keeling, W. F. Doolittle, and M. W. Gray. 2011. How a neutral evolutionary ratchet can build cellular complexity. *IUBMB Life* **63**: 528–37.

Lynch, M., and J. S. Conery. 2000. The evolutionary fate and consequence of duplicate genes. *Science* **290**: 1151–55.

Lynch, M., and A. Force. 2000. The probability of duplicate gene preservation by subfunctionalization. *Genetics* **154**: 459–73.

Marcus, J. M. 2005. A partial solution to the C-value paradox. Pp. 97–105 in A. McLysaght (ed.), *Ws on Comparative Genomics*. Springer, Berlin.

Maynard Smith, J., and E. Szathmáry. 1995. *The Major Transitions in Evolution*. Oxford University Press, Oxford.

McShea, D. W. 1991. Complexity and evolution: What everybody knows. *Biology and Philosophy* **6**: 303–24.

McShea, D. W. 1993. Evolutionary change in the morphological complexity of the mammalian vertebral column. *Evolution* **47**: 730–40.

McShea, D. W. 1994. Mechanisms of large-scale evolutionary trends. *Evolution* **48**: 1747–63.

McShea, D. W. 1996. Metazoan complexity and evolution: Is there a trend? *Evolution* **50**: 477–92.

McShea, D. W. 2001. The hierarchical structure of organisms: A scale and documentation of a trend in the maximum. *Paleobiology* **27**: 405–23.

McShea, D. W. 2002. A complexity drain on cells in the evolution of multicellularity. *Evolution* **56**: 441–52.

McShea, D. W. 2005. A universal generative tendency toward increased orga-nismal complexity. Pp. 435–53 in B. Hallgrimsson and B. Hall (eds.), *Variation: A Central Concept in Biology.* Elsevier Academic, Burlington, MA.

McShea, D. W. 2015. Three trends in the history of life: An evolutionary syndrome. *Evolutionary Biology* **43**: 531–42.

McShea, D. W., and E. P. Venit. 2001. What is a part? Pp. 259–84 in G. P. Wagner (ed.), *The Character Concept in Evolutionary Biology.* Academic Press, San Diego, CA.

McShea, D. W., and R. N. Brandon. 2010. *Biology's First Law.* University of Chicago Press, Chicago.

McShea, D. W., S. C. Wang, and R. N. Brandon. 2019. A quantitative formula-tion of biology's first law. *Evolution* **73**: 1101–15.

Mora, C., D. P. Tittensor, S. Adl, A. G. B. Simpson, and B. Worm. 2011. How many species are there on Earth and in the ocean. *PLoS Biology* **9**: e1001127.

Nabi, I. 1981. On the tendencies of motion. Pp. 123–27 in R. Levins and R. Lewontin, *The Dialectical Biologist.* Harvard University Press, Cambridge, MA.

Nuismer, S. L., and L. J. Harmon. 2015. Predicting rates of interspecific interaction from phylogenetic trees. *Ecology Letters* 18: 17–27.

Odling-Smee, F. J., K. Laland, and M. W. Feldman. 1996. Niche construction. *American Naturalist* **147**: 641–48.

O'Malley, M.A., J.G. Wideman, and I. Ruiz-Trillo. 2016. Losing complexity: The role of simplification in macroevolution. *Trends in Ecology and Evolution* 31: 608–621.

Penrose, O. 2005. *Foundations of Statistical Mechanics: A Deductive Treatment.* Dover, Mineola, NY.

Pfennig, D., and K. Pfennig. 2012. *Evolution's Wedge: Competition and the Origins of Diversity.* University of California Press, Berkeley.

Raup, D. M. 1977. Stochastic models in evolutionary paleontology. Pp. 59–78 in A. Hallam (ed.), *Patterns of Evolution as Illustrated by the Fossil Record.* Elsevier, Amsterdam.

Raup, D. M., and S. J. Gould. 1974. Stochastic simulation and evolution of morphology – towards a nomothetic paleontology. *Systematic Zoology* **23**: 305–22.

Revell, L. J., L. J. Harmon, and D. C. Collar. 2008. Phylogenetic signal, evolutionary process, and rate. *Systematic Biology* **57**: 591–601.

Roopnarine, P. D., G. Byars, and P. Fitzgerald. 1999. Anagenetic evolution, stratophenetic patterns, and random walk models. *Paleobiology* **25**: 41–57.

Ruse, M. 2009. *Monad to Man: The Concept of Progress in Evolutionary Biology*. Harvard University Press, Cambridge, MA.

Saunders, P. T., and M. W. Ho 1976. On the increase in complexity in evolution. *Journal of Theoretical Biology* **63**: 375–84.

Salmon, W. C. 1971. *Statistical Explanation and Statistical Relevance*. University of Pittsburgh Press, Pittsburgh, PA.

Salmon, W. C. 1984. *Scientific Explanation and the Causal Structure of the World*. Princeton University Press, Princeton, NJ.

Schopf, T. J. M, D. M. Raup, S. J. Gould, and D. S. Simberloff. 1975. Genomic versus morphologic rates of evolution: Influence of morphologic complexity. *Paleobiology* **1**: 63–70.

Sepkoski, J. J., Jr. 1978. A kinetic model of Phanerozoic taxonomic diversity. I. Analysis of marine orders. *Paleobiology* **4**: 223–51.

Sheets, H. D., and C. E. Mitchell. 2001. Why the null matters: Statistical tests, random walks and evolution. *Genetica* **112–13**: 105–25.

Sidor, C. A. 2001. Simplification as a trend in synapsid cranial evolution. *Evolution* **55**: 1419–42.

Stanley, S. M. 1973. An explanation for Cope's Rule. *Evolution* **27**: 1–16.

Stanley, S. M. 1975. A theory of evolution above the species level. *Proceedings of the National Academy of Sciences of the United States of America* **72**: 646–50.

Stanley, S. M. 1979. *Macroevolution: Pattern and Process*. Johns Hopkins University Press, Baltimore.

Sterelny, K. 1999. Bacteria at the high table. *Biology and Philosophy* **14**: 459–70.

Sterelny, K. 2016. Contingency and history. *Philosophy of Science* **83**: 521–39.

Stoltzfus, A. 1999. On the possibility of constructive neutral evolution. *Journal of Molecular Evolution* **49**: 169–81.

Valentine, J. W., A. G. Collins, and C. P. Meyer. 1994. Morphological complexity increase in metazoans. *Paleobiology* **20**: 131–42.

Vlieger, L. 2019. Book review – Biology's First Law: The Tendency for Diversity & Complexity to Increase in Evolutionary Systems. *The Inquisitive Biologist* (blog). https://inquisitivebiologist.wordpress.com/2019/07/23/book-review-biologys-first-law-the-tendency-for-diversity-complexity-to-increase-in-evolutionary-systems/.

Waddington, C. H. 1969. Paradigm for an evolutionary process. Pp. 106–28 in C. H. Waddington (ed.), *Towards a Theoretical Biology*, vol. 2. Edinburgh University Press, Edinburgh.

Williston, S. 1914. *Water Reptiles of the Past and Present*. University of Chicago Press, Chicago.

Acknowledgments

RNB and DWM thank Gene Hunt for contributing the data used in the demonstration in Section 4 and Steve Wang for help in developing the quantification of the ZFEL. DWM thanks Diane, Honey Bun, Trvsh, Sarah, and Naomi. Their contributions defy quantification. Thanks also to Benedikt Hallgrimsson for conversations decades ago which ultimately gave rise to the ZFEL for complexity. And thanks to you, Robert, for driving this, for – as you once put it – making me do it. With every one of our collaborations comes greater insight. RNB thanks Gloria and Katherine for support and Peanut and Theo for inspiration. And thanks to Dan for being the perfect collaborator, a collaborator who helped make this book be more than the mere sum of what each of us could have done by ourselves.

All or parts of the figures (except 3.4 and 4) appeared first in a paper in *Evolution* (McShea, D.W., S.C. Wang, and R.N. Brandon. 2019. A quantitative formulation of biology's first law. *Evolution*. 73(6): 1101–1115), and are reproduced here by permission of the journal, the Society for the Study of Evolution, and Wiley-Blackwell.

Philosophy of Biology

Grant Ramsey

KU Leuven

Grant Ramsey is a BOFZAP research professor at the Institute of Philosophy, KU Leuven, Belgium. His work centers on philosophical problems at the foundation of evolutionary biology. He has been awarded the Popper Prize twice for his work in this area. He also publishes in the philosophy of animal behavior, human nature and the moral emotions. He runs the Ramsey Lab (theramseylab.org), a highly collaborative research group focused on issues in the philosophy of the life sciences.

Michael Ruse

Florida State University

Michael Ruse is the Lucyle T. Werkmeister Professor of Philosophy and the Director of the Program in the History and Philosophy of Science at Florida State University. He is Professor Emeritus at the University of Guelph, in Ontario, Canada. He is a former Guggenheim fellow and Gifford lecturer. He is the author or editor of over sixty books, most recently *Darwinism as Religion: What Literature Tells Us about Evolution; On Purpose; The Problem of War: Darwinism, Christianity, and their Battle to Understand Human Conflict;* and *A Meaning to Life.*

About the Series

This Cambridge Elements series provides concise and structured introductions to all of the central topics in the philosophy of biology. Contributors to the series are cutting-edge researchers who offer balanced, comprehensive coverage of multiple perspectives, while also developing new ideas and arguments from a unique viewpoint.

Cambridge Elements ≡

Philosophy of Biology